The Nature of Plants

UNIVERSITY PRESS OF FLORIDA

Florida A&M University, Tallahassee
Florida Atlantic University, Boca Raton
Florida Gulf Coast University, Ft. Myers
Florida International University, Miami
Florida State University, Tallahassee
New College of Florida, Sarasota
University of Central Florida, Orlando
University of Florida, Gainesville
University of North Florida, Jacksonville
University of South Florida, Tampa
University of West Florida, Pensacola

THE NATURE OF

University Press of Florida

Gainesville

Tallahassee

Tampa

Boca Raton

Pensacola

Orlando

Miami

Jacksonville

Ft. Myers

Sarasota

PLANTS

An Introduction to How Plants Work

Craig N. Huegel

Frontispiece: *Passion Flower*, color drawing by Sarah Graziani.
Used with permission.

This book may be available in an electronic edition.

24 23 22 21 20 19   6 5 4 3 2 1

Library of Congress Control Number: 2018941382
ISBN 978-0-8130-6408-6

The University Press of Florida is the scholarly publishing agency for the State
University System of Florida, comprising Florida A&M University, Florida Atlantic
University, Florida Gulf Coast University, Florida International University, Florida
State University, New College of Florida, University of Central Florida, University
of Florida, University of North Florida, University of South Florida, and University
of West Florida.

University Press of Florida
2046 NE Waldo Road
Suite 2100
Gainesville, FL 32609
http://upress.ufl.edu

Contents

Preface and Acknowledgments

Plants have always been an important part of my life. Many of my earliest memories center on the garden, and it is a passion that has endured. I am a gardener first and a biologist second. I find plants to be fascinating, both for their beauty and their underlying biology. To me, these two aspects cannot be separated from each other. The beauty of plants is easily lost when they are distilled down to just the nuts and bolts of science, but understanding the basics of plant biology is necessary to fully appreciate what makes them special. My appreciation for plants has been greatly increased by understanding how they work, how they function internally, and how they relate to one another and the world around them, and all of this has made me a better gardener.

Gardeners do not need to know everything about the biology of plants to be effective, but understanding some of the basics is especially helpful. We need to know enough to understand what plants need to achieve their optimal growth, to reproduce, and to coexist with their neighbors and their enemies. Before we purchase a new plant to add to our landscape, we need to be able to honestly evaluate whether it is the right specimen to make the transition from potted plant to plant-community member. We need to understand how it will relate to the growing conditions of our landscape, and when something doesn't go right after we've planted it, we need to understand why. We don't have to become scientists to be good gardeners, but we should understand the science that's been produced by others. After all, we are dealing with finely tuned organisms that have been evolving intricate bells and whistles for nearly a half billion years.

The diversity of plants is astounding. Botanists estimate that there are nearly 288,000 unique species of plants in existence today. Nearly 90 percent

of these are flowering plants; the rest are mostly mosses, ferns, and conifers. Nearly every color, shape, and growth form exists to meet the aesthetic
needs of a modern-day gardener. This aspect is the one that calls most of us
to work with plants. It's the underlying biological part that has largely been
ignored.

Aristotle believed that plants had the lowest kind of "soul" because they
seemed incapable of any of the behaviors observed in animals. This attitude
has largely persisted in our collective psyche. Plants may be interesting to
look at, and they are definitely biologically necessary, but most of us view
them as somewhat boring, about as exciting as watching grass grow.

Many of us who garden often fall into this trap. We purchase plants, dig
holes, water them, and give them a jolt of fertilizer before largely leaving
them to live life on their own. If our plants fail to prosper, we water and
fertilize them again based on a belief that these simple solutions are all that
plants really require. If our plants die, we buy new ones or conclude that
our failure was somehow due to the plants themselves, and we then buy a
different kind to fill the void in our landscape. When we purchase or adopt
an animal for our household, we usually do things much differently. Prior
to bringing the animal home, we do some research on what our new addition will need from us and what we should expect from it, we take steps to
prepare our home to receive it, and we study its acclimation into our household—prepared to call in a veterinarian, a trainer, or some other expert
should anything appear out of order. After all, we may reason, animals are
complicated and plants are simple.

The truth is that plants are far more complex than we have traditionally
perceived them to be; in some ways they are even more complex than animals. Animals have life relatively easy. For one thing, they can move to more
favorable conditions when life treats them harshly. This ability to move gives
them lots of options that plants do not have. Once a plant's seed or spore
germinates, its only choice is to make do with what it's been given. Plants
must do everything animals do while remaining in the same location for
the duration of their lives. They have had to develop intricate adaptations
and behaviors over the more than 500 million years they have been on land.
That's a long track record of success. They seem simple to us because what
they do is largely done sight unseen and at a rate of speed that is hardly
noticeable.

This is not a book to teach you the fundamentals of plant biology, though it might do some of that in a roundabout way. Brian Capon has written a useful botany textbook for the general public (*Botany for Gardeners*, 1990), and there is no need to duplicate his effort. This book is intended to teach gardeners and others who love plants how to apply plant biology to make their work with plants more successful and to unlock some of the mysteries in a practical way so that more gardens thrive and fewer plants die. Along the way, I hope it also gives you a deeper appreciation for how intricate and interesting plants are. As a botany professor and a lifelong gardener, I want to better merge these two disciplines. Too many gardening books focus on what plants do for us, as if their sole purpose centered on their appearance and utility. My desire is that we look at plants without this anthropomorphic bias and see them for their own intrinsic value.

This book is an outgrowth of both my classroom and gardening experiences. As a plant biology professor, I routinely take complicated concepts and translate them for my nonbotanist students. I also try to focus on those things that I believe might be relevant to their future relationships with plants. My students are not interested in memorizing the intricacies of the Calvin cycle, for example, but they need to know how it works and why it's important. Very few of us, in fact, will ever be asked to sketch out the complex steps in photosynthesis, even though our lives depend on those chemical equations. Understanding the basics of how plants work is important, and that is the approach I have taken in the chapters that follow for all the aspects of plant biology that I've included. When we understand plants, we understand the foundation of life on earth.

I also wish to leave you with an appreciation for how plants work from the perspective of a fellow plant lover. I came to love plants for their beauty and their practicality. I have slowly evolved to love plants simply for their "plantness." Plants intrigue me because they are plants. I've come to look at them with a broader perspective than I once did. They have their own stories to tell. This book is my way of sharing their stories. I hope it will add to your love of plants and enhance your joy in working with them.

I wish to thank the many wonderful botany professors I've had over the years, both at the University of Wisconsin–Madison and Iowa State University. I feel fortunate that I attended those campuses when many of their most legendary instructors were there too. Since those days, I was given the opportunity to teach in the baccalaureate biology program at St. Petersburg College by Linda Gingerich. Her faith in my potential as an instructor has opened many doors to me, for which I am greatly indebted. The students who have passed through my classroom over the years have challenged me, required me to stay one step ahead, and generated countless hours of meaningful discussion. I hope that as they move forward in their lives they do so with a greater appreciation of plants and the roles plants play. They have been my test subjects in more ways than one.

My appreciation of plants and gardening has followed a long and somewhat circuitous path. I owe my earliest experiences to my father and mother, Jack and Louise Huegel, who always encouraged my curiosity about the living world. My growth as a gardener increased greatly since arriving in Florida because of the great many plant friends I've found here. Their names are too numerous to mention, but I hope I have made it clear to each that I value their time and friendship.

This book has been greatly enhanced by the work and creativity of others. I especially want to acknowledge those who provided the illustrations that grace these pages. My good friend Cathy Vogelsong took my general ideas and produced more than a dozen beautiful drawings. This book would not have been possible without her contributions. The cover art is the work of another good friend, Sarah Graziani. I suspect that many of you who first picked up and looked at this book did so because of her amazing artwork.

Although most of the photographs are mine, others provided some that I couldn't. My friends Christina Evans, Mira Janjus, Doreen S. Damm, Elizabeth Hoefler-Boing, and Juliet Rynear all contributed stunning photographs.

No book is written without the help of editors and careful reviewers, and this one is no exception. Early reviews were provided by Gil Nelson, Erin Goergen, and Nicole Pinson. Their input is greatly appreciated. As always, the staff at the University Press of Florida were invaluable in getting this book to print in the final form you see now. I am especially indebted to Linda Bathgate and Jane Pollack, and my primary editor, Susan Murray.

Gardening is more than a hobby to me, as it is to most of the real gardeners I know. For me, my plants are no different than my animal pets. I feel joy in their successes and sorrow in their failures. Though they may depend on me for their daily survival, I depend on them for mine. I do not know what my life would become without my plants and landscape to put-ter around in. It is my hope that these pages will add to your gardening joy and enhance the relationships you have with the plants you live with. As the French impressionist painter Claude Monet once wrote, "My garden is my most beautiful masterpiece." May all of our landscapes be works of enduring *and living* art.

Lichens are not plants but rather an association between a green algae and various fungi and/or cyanobacteria. They are the oldest photosynthetic organism to have colonized land.

Mosses and other bryophytes are the oldest true plants to have colonized land.

Modern bryophytes have evolved to meet the demands of a changing world, but they remain relatively simple compared to the vascular plants that evolved from them—the ferns, gymnosperms (conifers and their relatives), and flowering plants that are the focus of most gardening efforts. The evolution of vascular plants produced real roots and allowed for the development of complex above-ground structures because now there was a circulatory system that allowed for the transport of everything the plant needed throughout its entire body. This adaptation opened the door to many complex structures and behaviors, but it also allowed for the possibility of many things that could go wrong. Simple is sometimes just fine. Ask anyone who once worked on their own car before the advent of engine computers and electronics.

At their core, vascular plants have retained the simplicity of their ancient ancestors. They need sunlight to photosynthesize, and the basic C3 pathway that all nonvascular algal plants use still forms the core photosynthetic pathway of all their descendants. Though many higher plants have "improved" on this model and added a few bells and whistles to modernize it a bit, very little at the core of the process has changed. Plants need water to complete all the major functions necessary for life; they respire; they require certain elements to maintain a base metabolism; and when conditions are right, they grow, reproduce, and eventually die.

The bryophytes of 500 million years ago initiated an upward ladder of plant complexity starting with ferns. Ferns and their relatives (horsetails, whisk ferns, club mosses, spike mosses, and quillworts) were the dominant plants during the Carboniferous period, approximately 300 million years ago. Many of our richest coal deposits were produced by these "fern forests," and a huge diversity of species existed, including many as large as modern-day trees. Scientists believe that the earth was generally warmer and more humid during this time.

Ferns and their relatives made one especially significant change from their mossy ancestors: they developed a simple vascular system. By developing true roots, they could pull water, with all its dissolved nutrients, up from the soil or off a rock. The stems could then take the mineral-rich water and transport it into the fronds. Likewise, the chloroplasts inside the green fronds and stems could take the sun's energy, convert it to glucose, and then send those simple sugars throughout the entire plant. Unlike bryophytes, not every part of the plant needed to absorb water and perform photosyn-

Ferns and their relatives are the oldest plants to have developed true roots and vascular tissue.

thesis. They could have specialized parts in their overall structure. Ferns could now grow to be much larger than their mossy predecessors. All of this made for huge advances in the plant world, and it became the new model for every other type of plant to follow. From this point forward, all plants would have specialized roots, stems, and leaves.

Ferns and their relatives, however, still rely on relatively archaic methods for reproduction. Like the bryophytes that came before them, they produce spores instead of seeds, and they produce them in structures that rely solely on the wind to move them to new locations. To be fair, this was the most effective means of dispersal at the time of their evolution. There simply were no pollinators present in the Carboniferous period; there were plenty of insects, but there was nothing for them to pollinate. Wind has its limitations, especially in warm and humid air. Plants, like most of us, wish our children well as they move out into the world and become their own masters. We just don't wish for them to live at home their entire lives; nearby is fine, but on their own and independent of us. Windy days effectively move spores long distances, but plants have no control over the wind, and relying on chance is never the most efficient strategy for success.

Ferns and other fern-like plants maintained another characteristic of the bryophytes that also limited their reproduction. They required a film of water to get their gametes together for fertilization. Like the bryophytes, fern sperm is flagellated and can only reach a fern egg by swimming to it. They are completely dependent on the weather around them to produce new ferns. Reproduction cannot occur during periods of drought unless a layer of dew is produced overnight. More often than not, fern reproduction requires extended periods of moisture produced by rain and high humidity. This worked fine during the Carboniferous period but became more limiting as the earth's climate became less tropical. Today, most of these plants are found in tropical and semitropical areas or in the vicinity of wetlands or seasonally wet habitats. The incredible diversity of ferns and related plants found in the Carboniferous period has significantly declined.

Spores aren't all bad; they've worked successfully for hundreds of millions of years. Because they are so small, they can be produced in huge numbers with little investment on the part of the parents. Their minute size enables them to remain lightweight and capable of long-distance movements. They also can be quite resistant to temperature extremes and desiccation. The downside to spores is that they contain very little nutrition for the tiny embryonic plant inside. When a spore decides to sprout, the baby plant must find water and nutrients almost immediately if it is to survive its first days. There's little room for error. For that reason, the overwhelming number of spores produced by ferns and mosses perish before the young adult is even noticeable.

The evolution of seeds began in the Carboniferous period with the emergence of the so-called seed ferns, but it didn't really reach the advancement seen in more modern plants until the Permian period, approximately 50 million years later. By this time, plants easily recognizable as gymnosperms (conifers, cycads, gingkos, and gnetophytes) comprised the dominant plant life. This continued well through the age of the dinosaurs, ending approximately 100 million years later. Dinosaurs existed in a world largely devoid of flowering plants. The great herbivores so frequently shown in modern film and television feasted on pine trees, cycads, and the like—and did not encounter flowering plants until sometime late in the Cretaceous period, approximately 90 million years ago.

Gymnosperms perfected several significant new features that would become models for all the plants that would follow. Foremost was a reproductive system far less dependent on moisture and better designed to protect the developing embryo. Gymnosperms produced seeds in protective cones; they produced pollen instead of flagellated sperm to fertilize their eggs; and their fertilized seeds contained large stores of nutrients for the developing embryo inside. They could still cast huge quantities of pollen into the air because, like spores, there was little parental investment in producing it. However, they could focus more attention on the seeds, producing just a few in comparison to ferns and mosses but ensuring that a much higher percentage would actually develop and grow.

Seeds were a huge advancement in the evolution of modern plants. For one, the fertilized embryos are encased inside them. Instead of being encapsulated inside a thin casing as spores are, seeds have both a seed coat and a rich food source to protect and nourish the developing embryonic plant inside. Spores are cast out into the world by their parents with few resources on which to draw; seeds pamper the infant, protect and nourish it, and give it time to fully develop before it leaves the seed, sprouts, and becomes a functioning juvenile on its own. When the juvenile plant emerges from its seed, it is far more ready to face the world than any plant produced by a spore. It already has the beginnings of its roots and stems, it has embryonic leaves, and it often maintains a portion of its original nutrition stores to draw on while its roots fully develop their ability to draw water from the ground. Furthermore, its leaves have the ability to produce new food via photosynthesis. Baby plants produced by seeds start life with an advantage.

Producing seeds takes far more investment on the part of the parents, but it ensures a higher rate of survival for their offspring. The production of seeds also creates more opportunity to fine-tune the process of germination. Breaking the dormancy of a spore is a relatively simple process. Mostly it involves having the right moisture conditions around it. Spores can remain dormant for long periods of time waiting for just the right moisture conditions to occur; the embryo inside is in a state of suspended animation until then. Seeds, on the other hand, often have more elaborate mechanisms to prevent them from sprouting until everything seems just right. Seeds re-

Gymnosperms, like this cycad, are the first plants to have developed seeds as a means of reproduction.

quire some kind of stratification to break their dormancy: cold, heat, light, water, abrasion, or combinations of these conditions. By having a seed coat that requires certain conditions to be met before germination will occur, the embryo inside is protected from sprouting under conditions that are less than optimal for them to start life on their own.

Seeds also diversify dispersal methods. Whereas the simple rounded spores of mosses and ferns require an air current to move them away from their parents, wind-dispersed seeds are often aided by the presence of "wings" to help carry them farther. This was a major advancement, but most seed-producing plants completely abandoned their reliance on wind. They developed other options, and the most significant new vector was movement by animals. When swallowed whole, seeds pass through the digestive tract of animals and are released in their waste as they move about. Animals that make long-distance movements can carry seeds to completely new habitats not already colonized, and the seeds are deposited with their own dollop of fertilizer. The gymnosperms initiated this mode of dispersal, but it was the flowering plants that developed later that perfected it.

Gymnosperms were also the first plants to truly protect their offspring during early development. They did this by producing their eggs inside rigid cones and then surrounding them inside these cones until the embryos were fully developed. While "seed ferns" merely held their primitive seeds inside open, cup-like structures, gymnosperms encased them in woody, often spiny, scales. In modern-day pines, for example, it can take more than a year for fertilization to occur once the pollen is captured inside the egg-producing cone. Then it may take another twelve to sixteen months for the seeds to develop and mature. For all this time, everything is well protected inside the cone. The spores of ferns and mosses are in no way so well guarded.

Last, gymnosperms do not require water for fertilization to take place. Their sperm are encased inside the pollen grains, and the pollen is dispersed in the wind. Like spores, pollen is lightweight and capable of long-distance movement under the right environmental conditions. The pollen of gymnosperms comes surrounded by two "wings" for extra buoyancy in their aerial travels, but the sperm inside does not require a film of water on which to travel to the egg. Some primitive gymnosperms still have flagellated sperm that evidence their evolutionary connection to the ferns and mosses, but they don't use them in fertilization. Their pollen actually creates a tube that burrows toward the unfertilized egg and deposits the sperm next to it. The evolution of gymnosperms created an intricate new level of complexity not seen in plants previously. Reproduction was fine-tuned to increase the likelihood of success. Sperm now had to have a burrow created for it to reach the unfertilized egg. Females now had more control over which sperm would do the fertilization, and the entire process was less haphazard.

Some gymnosperms (the conifers and gingkos) also evolved the ability to produce wood and thereby increase their girth. True wood did not exist in the tree ferns or any of the other fern-like plants that composed the forests of the Carboniferous period. Wood and increased girth allowed for upward growth not previously possible. Pines (*Pinus* spp.), cedars (*Juniperus* spp.), spruces (*Picea* spp.), cypress (*Taxodium* spp.), and the like can reach mature heights of more than 100 feet—something previously impossible. Mature redwoods (*Sequoia sempervirens*) and sequoias (*Sequoiadendron giganteum*) routinely exceed 350 feet. Wood permits a great many functions. It provides structure, allows other plants to attach and grow, and adds structural complexity to plant communities.

Gymnosperms dominated the earth's upland habitats for at least 150 million years, sharing the area with dinosaurs until late in the Cretaceous period—approximately 100 million years ago. Relatively recent information gleaned from the fossil record demonstrates that flowering plants emerged in the prehistoric flora well before an asteroid struck the Yucatán Peninsula and changed everything with a global mass extinction event. The large-scale habitat changes that followed this asteroid strike allowed the remaining plants to quickly evolve, and what evolved were flowering plants. More than 350,000 species of flowering plants have emerged in the last 100 million years, more than all other plant types combined. Our modern world is completely dominated by plants that produce flowers.

Flowers provided plants with an ability no other plant possessed: a means of pollination not reliant on wind power but rather facilitated by the movement of pollinating animals. Wind had worked rather efficiently up to this time, but using animals to transfer seeds and pollen opened a new door that has proven extremely effective. With the evolution of flowers came an explosion in the evolution of pollinators, especially among

The success of flowering plants is largely the result of their association with pollinators such as bees and butterflies.

pollinating insects. Modern-day bees and pollinating wasps evolved from predatory species. Prior to flowers, there was no need to collect, feed on, and disseminate pollen. Evolutionary ecologists believe that butterflies evolved from ancient caddis flies (the group that includes bagworms) at the same time flowering plants began to proliferate. There is no doubt that the evolution of flowers coincided with the evolution of pollinators and that this relationship continues and defines many of our modern gardening practices.

Pollinators allowed for selective pollination; plants could develop complex relationships with their pollinators and pollinators could become extremely specialized. The element of chance that was the driving force behind wind pollination was largely circumvented. Of course, all of this meant that plants became ever more complex, and complexity brings both benefits and difficulties. Suffice it to say that since nearly 85 percent of the world's plants produce flowers and nearly 85 percent of those are animal-pollinated, this is a very effective strategy.

Pollinated flowers produce fruit, and that too was a major advance over the cones produced by gymnosperms. While gymnosperms produce seeds protected beneath the covering of a hardened scale, the seeds themselves are mostly reliant on wind for dispersal. There are some exceptions, but pines, spruces, and others in the conifer family produce seeds covered by a light diaphanous wing. When the seeds are ripe, the scales of the cone open, releasing the winged seeds to float away in the wind. Flowering plants such as the grasses and orchids still rely on wind power to disperse their seeds, but the majority of flowering plants use animals. The evolution of fruit coincides with the evolution of fruit-eating animals (frugivores). Producing succulent, high-energy fruit like apples (*Malus pumila*) and peaches (*Prunus persica*) takes a lot of energy on the part of the parent plant. That expenditure of energy is justified when it serves to attract animals that eat the fruit and carry the seeds to distant places before depositing them. Just as flowers and pollinators have been able to develop complex relationships over time, so have fruit-bearing plants and frugivores. Not all fruit is equally attractive to every animal. Flowering plants' ability to target specific groups of animals with their fruit meant that the plants could become extremely "discerning" in the ways seeds would be dispersed.

Fruit is not only supremely edible, but, in many cases, it is also extremely good at protecting the developing embryos inside. In fact, the relationship between those embryos and the surrounding fruit is quite complex. As the embryos develop, they release a suite of hormones that influence the development and ripening process of the fruit itself. When the embryos are young, the surrounding fruit is typically hard, bitter, and largely inedible. In this condition, the seeds enclosed inside are protected from danger. As the embryos inside the seeds get closer to maturity, however, it is in the plants' best interest to speed the ripening of their fruit and increase its palatability. The seeds of a green tomato (*Solanum lycopersicum*) are not mature enough to sprout and grow into new tomato plants; most animals wait until the fruit is at the peak of ripeness before consuming it. In this way, the fruit of most flowering plants is both a protective vessel and a dispersal mechanism.

We garden with plants that have been evolving for hundreds of millions of years. At each step they have taken up the evolutionary ladder, from bryophytes to modern-day flowering plants, their complexity has increased. For the most part, plants can take care of themselves. They don't really need us to coddle them. At the same time, we can't pretend that they are simple. Plants challenge us to understand them if we are going to treat them respectfully and ultimately carve out a space in our landscape that will satisfy both of us.

1

Light

THE MIRACLE OF PHOTOSYNTHESIS

Well before plants, the earth's oceans teemed with organisms that had evolved the ability to capture the energy of the sun and convert it to a simple six-carbon sugar known as glucose. It is generally agreed that the first organisms capable of photosynthesis evolved more than 3 billion years ago among certain types of bacteria. Some of these, like the cyanobacteria, also developed the ability to convert atmospheric nitrogen to forms useful to cellular functions like protein production. When people warn me about the dangers of bacteria, I have to remind them that all life on earth owes its very existence to the presence of bacteria despite the few that cause disease.

The evolution of photosynthetic pathways that used water and produced oxygen as a waste product came a bit later; the date is the subject of much discussion among the experts. Several different lines of geological evidence indicate, however, that oxygen began to accumulate in the atmosphere about 2.4 billion years ago. This suggests that the modern photosynthetic pathways used by plants and their ancestors evolved much earlier. Atmospheric oxygen can't increase without a coetaneous decrease in CO_2, so we know that the sequestration of carbon that led to our modern-day fossil-fuel industry began several billion years ago.

Life as we know it has been made possible by the evolution of photosynthesis. Though there are various specialized organisms that don't deploy that process, animal life is dependent on plants, and plants are dependent on photosynthesis for every aspect of their life. This miracle of life is quite simplistic on the surface. Though the various steps inside the equation get rather complex, the overall equation is easily comprehended:

$$6 \; CO_2 \; \text{(carbon dioxide)} + 12 \; H_2O \; \text{(water)} + \text{sunlight} =$$
$$C_6H_{12}O_6 \; \text{(glucose)} + 6 \; H_2O + 6 \; O_2 \; \text{(oxygen)}$$

The carbon dioxide plants need for photosynthesis comes from the atmosphere. It enters the plant through pores in the leaves and green stems called stomata. Water enters the plant through its roots (unless it is a moss or other bryophyte). The photosynthetic process takes place during daylight hours (with a few exceptions) as sunlight strikes the plant. Plants take the carbon dioxide existing in the atmosphere and the water that they already contain in their cells and use the energy of the sun to split the carbon dioxide into free oxygen while taking the now-free carbon to make glucose. The oxygen so vital to plant and animal respiration is actually a by-product of the plant's need for carbon to survive and grow. Solar energy is converted to chemical energy, and this chemical energy fuels all higher plant and animal life on earth. When someone tells you that sugar is bad for you, realize that all life on earth is dependent on the sugar produced by photosynthetic plants. It's the other types of sugars and how we use them that are negatively affecting our health.

Photosynthesis is possible because of a complex molecule known as chlorophyll α. Chlorophyll α actually starts the photosynthetic process. It is the molecule that sunlight has to hit in order to split carbon dioxide into oxygen and carbon molecules, and it's the site that then takes this extra solar energy and eventually stores it in the chemical bonds of simple and complex sugars. When we talk of getting a "sugar high" from eating too many sweets, we are really getting that "high" from sunlight.

Though other chlorophylls have been found in plants, they merely supplement the ability of chlorophyll α to perform its job. All plants (and their precursors) have taken this chlorophyll molecule and packed it into storage bodies known as chloroplasts. Each chloroplast contains many flattened disks known as thylakoids in which the chlorophyll is stored, and these thylakoids are stacked into cylindrical silos known as grana. Grana are then interconnected into distribution centers known as antenna complexes. The material surrounding the grana is called the stroma. Light strikes the chlorophyll in the chloroplasts and initiates photosynthesis. Oxygen is released from the initial reactions in the thylakoids (the Calvin cycle, or C3 pathway); glucose is produced in the second part of the reaction in the stroma. Despite all the complexities of turning solar energy into carbohydrates, the actual structure that produces it is rather simple—thousands of molecules of chlorophyll α, collected into hundreds of flattened disks, stacked into tens of cylinders and then interconnected with one another within each chloroplast and surrounded by a matrix of stroma. Every cell in a green leaf or stem may have a dozen chloroplasts. The total number of chloroplasts in even a small photosynthetic plant is mind-boggling. In typical plants, the chloroplasts required for photosynthesis are concentrated just below the leaf and green stem surfaces. These so-called C3 plants use an enzyme known as RuBisCo to initiate the photosynthetic process. The evolution of RuBisCo and this system of chlorophyll, chloroplasts, thylakoids, grana, antenna complexes, and stroma has worked for millions of years without any real modifications.

You don't need to know the many intermediate steps of photosynthesis to grasp the overall picture. The important message is that sunlight provides the energy that is needed, and carbon dioxide provides the oxygen required for plant and animal respiration and the carbon required for plant growth. Without water, none of this would be possible. Life as we generally know it is run on the energy supplied by the sun and harnessed by the ultimate solar panels in the universe—located in chloroplasts.

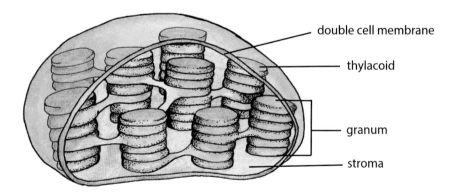

Detail of a chloroplast.

Sunlight strikes the earth as photons—packets of electromagnetic energy whose total energy depends on its wavelength. Waves with lower amplitudes and wider spacing between their crests carry far less energy than the taller, closely spaced ones. You see the same effect if you spend a day at the beach. It is much easier to stand in the water when the waves striking you are slow and rolling than after a speeding boat races by and magnifies the wave heights. You can easily feel the energy difference between these two types of waves. Energy emitted by the sun is composed of a broad spectrum of wavelengths, and all life on earth is constantly being struck by all of them. The wavelengths somewhere in the middle of this broad spectrum are known as "visible" light. It is the portion we see with our eyes. We see it as a rainbow of colors when it is passed through a prism or through water droplets produced in a rainstorm. Visible light is also the portion of the spectrum that plants use for photosynthesis. In a way, we have evolved together.

Other portions of the solar spectrum are not useful to this process. The sun emits longer wavelengths in the form of infrared radiation that some insects and birds can see, and even longer wavelengths in the form of radar, AM and FM radio, television, and A/C circuit waves. These lower-energy electromagnetic solar waves cannot be seen by us or used by plants for photosynthesis, but they can be felt as warmth. The thermal portion of sunlight comes from infrared wavelengths striking the earth's surface. The longer, lower-amplitude waves pass more easily through clouds in the atmosphere than do the higher wavelengths. Therefore, we can feel warm even on a cloudy day. Because these wavelengths are lower in energy, however, they do not carry the energy needed to initiate photosynthesis. Visible light packs enough of a wallop when it hits a chlorophyll molecule to split the carbon dioxide and water molecules necessary to start the process. Infrared and other less energetic wavelengths just can't do it. They warm the plant, but they do not provide it with the energy needed to survive.

The solar-generated electromagnetic spectrum also contains wavelengths with more energy than that contained in visible light. Ultraviolet radiation, X-rays, and gamma rays constantly pass through the earth's atmosphere too. We don't see them either, but we see and feel the effects of their presence when we get sunburned after spending time outdoors without protection.

Clouds deflect some of these rays, so we are less likely to burn when it is overcast than when it is not, but it is just as easy to burn on a cold day as it is on a warm one. Too much exposure to these high-intensity rays is damaging to our health.

Ultraviolet light, X-rays, and gamma rays are too energetic to work with chlorophyll molecules in photosynthesis and can damage plant tissues in the same way that they damage ours. We cover ourselves in lightweight clothes, broad-brimmed hats, and sunscreen if we are prudent when we spend time working outdoors under the full sun. Plants do much the same thing if they are adapted to thrive in sunny locations. They cover their leaves with hairs (trichomes) and waxy cuticles, and they produce carotenoid pigments to serve as a sort of sunscreen. Carotenoids are the orange, red, and yellow pigments that become apparent in the fall as leaves begin to turn color. Actually, they are present at all times; their colors are masked during the growing season by the chlorophyll. Carotenoids cannot fully protect plants that are poorly adapted to large amounts of ultraviolet light. We receive sunburns and turn red; plants turn a sickly yellow as their chlorophyll gets destroyed faster than they can manufacture more.

Though sunlight contains a very broad spectrum of wavelengths, only the relatively narrow band that we call visible light is useful to photosynthesis. Even within this narrow band, however, plants are selective. Not all the colors of the rainbow are equally effective, and some portions are not used at all.

Chlorophyll α, the molecule solely responsible for driving the photosynthetic process, reacts primarily to visible light in the orange/red portion of the spectrum and less so in the area at the division between violet and blue. It does not respond to the part of spectrum that includes green and yellow. If you were to grow a plant using a specialized bulb or filter that included only this portion of the visible spectrum, your plants would not be able to photosynthesize and would quickly die just as if they had been kept in the dark. In fact, we think of plants and chlorophyll as being green in color when in fact they are green only because they do not absorb the green portion of the visible spectrum. They reflect it back to our eyes. Chlorophyll is not a green pigment; it is a green light reflector.

While chlorophyll α drives the process of photosynthesis, additional chlorophylls and pigments supplement the process. To date, scientists

Though sunlight contains a vast spectrum of energy, plants use only the portion known as visible light—the colors found in a rainbow. Photograph by Christina Evans.

have found three other chlorophylls, essentially identical to chlorophyll α except for small changes to a single side chain in its molecular structure. Of these, only chlorophyll β occurs in true plants with chlorophyll α. The evolution of a second type of chlorophyll has allowed plants to capture a bit more of the visible light spectrum and use it in photosynthesis. Chlorophyll β is able to use the energy of a slightly different part of the blue and orange/red spectrum and pass it on to chlorophyll α to make it more efficient.

Additionally, the carotenoid pigments that double as a sunscreen also capture parts of the blue spectrum not captured by either of the chlorophylls. Like chlorophyll β, they take this energy, pass it on to chlorophyll α, and increase its efficiency. To summarize, plants are capable of photosynthesizing only within the red/orange and violet/blue segments of the visible light spectrum. Specialized bulbs for use in indoor situations take advantage of this and maximize light in these areas while minimizing light in the green/yellow range.

Not all light is created equal. Light comes as wavelengths with different energy signatures, and only a very narrow part of the light generated by the sun fits the needs of a plant's chlorophyll molecule to generate the complex processes of photosynthesis. It is a lock-and-key approach, and the key is highly specialized. After several billion years of evolution, we shouldn't expect anything less intricate. The process of photosynthesis is indeed a miracle.

Plants feed on light and use it to grow. Therefore, light (or more accurately, a very narrow portion of it) is plant food. Fertilizers aren't. When we fertilize our plants, it is equivalent to humans taking a multivitamin. A good fertilizer supplies the elements plants need to carry on the many enzymatic reactions involved with photosynthesis or to keep their cells functioning in top condition. Sunlight, however, provides the carbon molecules that produce simple sugars, which eventually get made into more plant tissue. New plant tissue is produced by the sun and the CO_2 in the atmosphere.

When we understand this, we can better grasp the very real importance of providing our plants with the diet of sunlight they need to remain healthy. Just like animals, plants have baseline metabolisms. Some require large amounts of sunlight to function properly, and others do just fine with less. The plants of open pastures and deserts, for example, have evolved to live under full sunlight. They consume large quantities every day, and they begin to decline when given less. Plants adapted to grow in the understory of dense forests, however, adapt by slowing their metabolisms down and feeding far less ravenously on sunlight to stay fit. While some plants can adjust to slightly more or less sunlight than their optimum, most can't, and their decline is either slow and hardly noticeable or quick and obvious. If you put a moth orchid (*Phalaenopsis* spp.) or a begonia (*Begonia* spp.) under full sun, its leaves will soon scorch, and its death will swiftly follow. If you subject your turfgrass to the dense shade of a large canopy tree, it will eventually weaken and die, regardless of how much water and fertilizer you apply to it. Carbon is the building block of life, and it comes from the atmosphere, harnessed by photosynthesis.

Like animals, plants use their food to support every aspect of their life requirements, and they dole it out for their various needs based on a hierarchy of importance. First, they use the sunlight they receive to support their core metabolism. After that, they use it to grow and then to reproduce. That

Too much sun can overwhelm a plant's natural ability to protect itself and can scorch its leaves.

simply makes sense. Here is an apt analogy: If you've ever lived in subzero temperatures, you understand frostbite. As your exposure to cold increases, the first symptoms appear in the appendages and extremities—fingers, toes, ears, and the tip of your nose. Such things are not necessary to preserve your life. Your brain and heart—which are important—are the last to go before death. Plants function the same way when confronted with inadequate light. When given a starvation diet of light, they will maintain their core life functions and give up the less necessary ones.

Confronted with inadequate light conditions, plants will first abandon their inclination to reproduce. Why bring children into a world with no resources for them? Flowering and fruiting are the least of a plant's worries when light is limited. Even so-called shade-tolerant plants need sufficient light to flower, and a lack of flowering is often the first indication that you are light-starving your plants. I have made this mistake so often in my own gardening experience that I've lost count. Perhaps the best recent example is with a small cluster of sweet shrub (*Calycanthus floridus*) that I planted more than twenty years ago beneath the shade of two deciduous canopy trees. Where I live in central Florida, sweet shrub is about 150 miles south of its natural range, so I believed I was making conditions easier for it by protecting it from our more intense sun. I was not. For more than two decades, these sweet shrubs went through their annual cycle of leafing out in the spring and going deciduous in the winter. They even grew a bit, but they never flowered. During these years, I attributed all of this to the fact that our winters were not cold enough to stimulate flowering, but then the canopy trees were taken down for other horticultural reasons. The next spring, the tiny patch of sweet shrub burst into bloom. Its diet of sunlight had been increased, and it now had enough energy to do more than simply survive.

The second thing a light-starved plant will do is fail to grow. If the most important thing is to preserve its core metabolism and survive, there is no sense in becoming larger and requiring more energy. A plant will first give up sex. Then it will fail to grow. Though growing toward the light may eventually allow it to increase its intake of sunlight, it still requires energy to grow that way, and this tradeoff is seen every day in most landscapes. If a plant has enough light to satisfy its core metabolic needs, it can use the extra for growth. If there is more sunlight above it in the canopy, it will grow taller to reach it, even if that means sacrificing girth and strength. If the nearest source of light is to one side or the other, it will grow sideways to reach it, even if that makes it unbalanced and more prone to toppling over. When we place plants in inadequate light, they will try to correct the problem if they can. If the light is too inadequate, however, they will simply shut down and not get larger. I have three fringe trees (*Chionanthus virginicus*) in my landscape, the last two purchased at the same time and of the same size when planted almost a decade ago. I planted both in the backyard, but one was in a sizeable light gap, and the other was deeper into the canopy of my existing "woodland." The shade-planted specimen bloomed the year I planted it but has never done so since, nor has it grown. This past year, it lost nearly all its main stem and sprouted weakly from the bottom of the trunk. It is slowly dying, but I have no place to move it to. The one planted in the sunnier spot blooms profusely each spring and has more than tripled in size.

Fertilizer and water will not correct inadequate light. Photosynthesis requires water and CO_2; both occur equally in sunny and shady landscapes, all else being equal. What is not present is the required energy to

This blooming fringe tree (*Chionanthus virginicus*) was planted in a location where it receives optimal sunlight.

drive the photosynthetic process, the "food" that feeds the plants. Plants that fare well in low light generally do so by evolving a much lower metabolism than others. They are slow growers, and they use the energy they receive sparingly. Most do not flower profusely or produce a lot of fruit. Some do so only in alternate years or even more sporadically than that. Some have adapted so well to a low-light diet that they can't handle higher levels. The stimulation of higher light levels is actually lethal. Other plants simply make the adjustment. They survive and flower and grow, but they do so at a much slower rate that fits their lower-light diet. Our Simpson's stoppers (*Myrcianthes fragrans*) are a classic example of the ability some plants have to adjust. In its typical south Florida habitat, Simpson's stopper is an understory subcanopy tree, and it resides there beneath an evergreen overstory that shades it all year. The two Simpson's stoppers planted in our wooded backyard are beautiful specimens, but they produce few flowers or fruit. The half dozen planted along the street as a screen flower profusely twice a year and produce copious amounts of fruit.

Plants can survive inadequate sunlight, but they can't reach their full potential in such conditions. They also can't store energy to use when faced with a catastrophic situation such as losing their stem to storm damage or major herbivory. They are living life at the "edge." Gardening is not about survival but about giving your plants the conditions where they can reach their full potential. Over my lifetime, I have planted far too many plants in settings where they have merely survived, and I have waited far too many times for them to someday "catch on." No plant will someday start to perform adequately if it has been placed in a location bereft of sufficient sunlight to meet its full range of life requirements. Feed your plants with a diet of sunlight, and they will prosper. If you can, create more light by selective pruning to open the canopy; if that is not possible, move them to a new location. Last, do not prune away a significant portion of their photosynthetic engine (their leaves and green stems) if you expect them to grow to their full potential.

PHOTORECEPTION

Most of us would be lost without an ability to know the time of day. We keep all kinds of instruments in our possession to inform us of the time, and we schedule most of our daily activities based on the surety that we will know exactly what time it is at any moment. For all of time's importance to us, it is even more important to plants. After all, if our clocks are off by a few minutes or hours, it is rarely a life-or-death situation. Not so with plants, and because of that, they have developed rather precise time-reading abilities. They achieve this by "reading" light—both its intensity and the change in day length.

The vast majority of decisions that plants make in their lives—from knowing when to flower or lose their leaves to where to turn their flower heads before the sun rises in the morning—is based on light and the ability to sense it. Higher animals use their eyes to make these decisions, but

Plants grow toward the light and retain their form as they do so. This "lucky bamboo" does not grow this way naturally but was carefully manipulated over time by turning it toward a light source.

light to begin with—the "shade-tolerant" species—or to make sure that they receive adequate light throughout the day, not just in the early morning or late afternoon, for example. This is easy when creating a canopy where none existed formerly. By planting the canopy at the same time, these trees will have equal shares of sunlight as they mature and their respective canopies

fully form. If planting under an existing canopy, use judicious canopy pruning techniques to create overhead light gaps instead of sideways ones.

Phototropins, with their blue-light-sensitive pigments, do much more
than direct plants toward light. As stated above, they assist in photosynthesis. Under low-light conditions, phototropins alter the position of the
chloroplasts so that they are perpendicular to the incoming light. In this
way, they optimize light absorption. In intense sunlight, however, they do
the opposite; the chloroplasts move away from the most irradiated edges,
and this protects them from damage. Though carotenoids serve as a sort of
sunscreen, the phototropins act like your parent who simply tells you that
enough is enough and that you are coming inside regardless of how much
sunscreen you have applied.

Phototropins also open the stomata that occur on leaves and herbaceous
stems, allowing CO_2 inside for photosynthesis. Understandably, the timing
of this process is critical to the plant's well-being. Stomata must open for
photosynthesis to proceed. Except for a few very specialized desert plants
that open their stomata only at night to conserve water, plants open their
stomata in conjunction with the hours of daylight and close them when
photosynthesis is impossible at night. Such sensitivity requires plants to
have a finely tuned sensory system, one dependent on phototropins.

Phototropins are involved with the synthesis of chlorophyll, the molecule
at the heart of photosynthesis and absolutely vital for all aspects of a plant's
growth and development. Chlorophyll molecules, however, are constantly
degrading over time and need to be regenerated. Phototropins, triggered by
blue light, are a key component of that regeneration process.

The other set of pigments, the phytochromes, use light within the red
portion of the spectrum. Phytochromes regulate the important functions
related to the inducement of flowering, chloroplast development (not including chlorophyll synthesis), leaf aging, and leaf fall. These functions are
collectively known as photoperiodism. Phytochrome pigments sense the
changing patterns of day length throughout the year and control a great
many plant behaviors and functions other than growth and photosynthesis.
Flowering plants are largely sensitive to day length. Most of the world's flowering plants are divided up into three main groups: those that flower after a
period of long nights; those that flower after a period of increasingly short
nights; and those largely unaffected by day length. The day-length-neutral

plants mostly come from tropical locations where day length is constant throughout the seasons. In other latitudes, it is crucial that plants attempt reproduction at precisely the right time of year. Though plant biologists often refer to these as either "long-day" or "short-day" plants, it is really the length of darkness to which plants are responding. Long-day plants sense an ever-decreasing length of night as spring gives way to summer. Such species produce their flowers at these times of year and set seed before winter. Short-day plants, on the other hand, sense that the hours of darkness are increasing. They flower in spring.

For this to happen effectively, such plants must be especially sensitive to the length of daylight they are receiving. They have to collate this with other weather factors such as temperature, and they are triggered into an action once a certain threshold is reached. They can't be fooled by periods

Many plants time their flowering using day length as a guide. Christmas cacti (*Schlumbergera* spp.), like this one, are triggered to bloom by the shortening length of sunlight from fall to winter.

of clouds, for example. Clouds affect the intensity of sunlight, not the length; therefore, weather patterns like this are mostly irrelevant. What can affect this delicate system, however, are things that interrupt plants' ability to detect patterns and thresholds of darkness. The length of darkness is what plants sense, not the length of light. Interruptions to that unbroken length of darkness can have very visible impacts.

Horticulturalists use this sometimes to "trick" a plant into bloom at a time of year that is not natural to it. Poinsettias (*Euphorbia pulcherrima*), for example, need to bloom at the Christmas holidays to sell. If they flowered in July, almost no one would purchase one. The fact that millions of poinsettias come to bloom each year at exactly the "right" time is because they are grown under a controlled regimen of light and dark in a greenhouse.

Sometimes we interrupt this system inadvertently. Night lighting is one of the most common inadvertent impacts of this type. While bursts of darkness in the daytime would go unnoticed by flowering plants, bursts of light or continuous light during periods of darkness can completely overwhelm a plant's ability to use its innate photoperiodism to produce buds and eventually flower. The intensity of the light and the frequency of the light waves emitted by the bulbs are especially important.

A plant's phytochrome pigments will not react to light of low intensity. There simply is not enough energy to trigger a response in such situations. You can tour your landscape with a flashlight every evening, for example, and nothing will happen to your plants because of it. Low-wattage bulbs or bulbs set away from the plants will not disrupt their ability to flower, but bulbs set within the actual landscape, especially those of high intensity, may. When nontropical plants cannot accurately measure the length of darkness, they may fail to flower completely or flower out of season.

Photoperiodism influences many activities besides the onset of flowering. Phytochromes affect growth, seed germination, fruit development, and the onset of winter dormancy—things that are key to ecological success. Growth, for example, should occur in the seasons best suited for it and decline or stop at other times. Long periods of light (actually short periods of darkness) provide information to a plant, irrespective of temperature, that photosynthesis is highly ramped up and that a lot of this energy should go into growth. The same is true for winter dormancy. Winter-dormant plants would be taking a significant risk if they used temperature alone to tell them that spring has arrived and that it is time to leaf out. Phytochromes communicate to the plant that spring has only really arrived when the hours of darkness have been reduced by a certain threshold number. As I write this in mid-December, we have had weeks of high temperatures in the mid-eighties (Fahrenheit) at my central Florida home. Though some of my plants have leafed out sporadically in the seasonally warm temperatures, most have not. Their phytochromes are telling them to wait for a few more months.

It also benefits most plants that seed germination and fruit development are regulated by phytochromes. This is not a concern for tropical species, but those in more temperate zones require accurate information on the progress of the seasons independent of temperature. Fruit needs to develop at the time of year best suited to each plant's ecology, and the infant plants, nestled inside each developing seed, need to sprout when it is safest. If every winter warm spell caused the seeds of temperate and arctic plants to sprout, the vast majority would die with a return to "normal" winter conditions. The embryo inside the seed is a baby plant, and it produces a wealth of hormones to ensure its survival. Some of these are triggered by phytochromes.

Phytochromes are the pigments most involved in maintaining a plant's biological clock. Not only do plants need to measure the changing seasons through seasonal changes in day length, but they also need to sense the twenty-four-hour day and adjust their activities accordingly. We've understood the role this plays in animals for some time. Plants as well as animals maintain an internal biological clock, known as "circadian rhythm," that helps them determine the passage of twenty-four-hour blocks of time. This biological clock will work for days even when plants are completely deprived of light. A biological clock ensures that certain functions performed by the plant are maintained uninterrupted even when outside cues are lacking for a time. Eventually, this clock will start to fail in total darkness and needs to be reset by time in the sun.

Circadian rhythms are often referred to as "sleep movements." Many plants adjust the position of their leaves and flowers during the evening and readjust their position just before daybreak to take full advantage of sunrise. Plants relax their leaves at night to fold or droop when there is no reason to maximize their surface area for photosynthesis, but they extend them

The brilliant fall colors that we admire occur in response to the complicated relationship between shortened day length and cooler temperatures. As plants assimilate this information, they quit producing chlorophyll and prepare to drop their leaves for winter.

outward at 90-degree angles again in time for another day. Many also droop their flower heads at night, especially those that are pollinated by daytime pollinators. Plants such as sunflowers (*Helianthus* spp.) can actually predict the direction of the sun and readjust their flower heads to make maximum use of the sun once it rises above the horizon.

Many plants exhibit outward responses to day and night, known as sleep movements. This wood sorrel (*Oxalis debilis*) opens its flowers each morning and closes them at night when its pollinators are asleep.

Many flowering plants adjust the scent of their blooms based on this circadian rhythm. I have several night-fragrant orchids at my home. These, like many other flowering plants, maintain their open blooms for five to six days in a row, but they are fragrant only at night to attract the moths that pollinate them. They maintain this pattern even if brought inside during their flowering period. It is controlled by their biological clock. Likewise, a great many other flowering plants are fragrant only during daytime hours.

Last, some plants that have flowers that open for only one day are exceptionally specific as to when they open. I have two of these in my home landscape. Bartram's ixia (*Calydorea caelestina*), a distant relative of the irises, times its spring blooms so that they open almost immediately when greeted by the morning sunlight, but they remain open for only a few hours and close well before noon. Fall ixia (*Nemastylis floridana*), on the other hand, opens its flowers only in midafternoon and closes them several hours later, well before nightfall. Photoperiodism is set by phytochromes and triggered by red light, but it is also influenced by temperature.

When plants get conflicting cues, it can throw off their normal patterns. Here in central Florida, it is not unusual to have warm falls that turn into warm winters. Based on changing photoperiods, many of my deciduous plants "know" it is fall and lose their leaves at the appropriate time. Others, however, will hold their leaves weeks later than is typical and never produce fall color before leaf drop. As I write this, our unusually warm November–December has played havoc with a small subsample of my north Florida native plants; these are now leafing out again and producing a few flowers. Photoperiodism is not the only controlling factor, and when temperatures are not in balance with the season, they may pull a plant in two separate directions.

A plant's response to light is also controlled by its genetics. This is not especially apparent in plants with rather narrow geographical ranges or in tropical or semitropical species, but it's extremely obvious in species such as red maples (*Acer rubrum*) with extensive ranges. Red maple occurs naturally from northern Canada to the Everglades in southern Florida. Across this range, individual populations have evolved to time their activities according to the regional climate. Therefore, red maples from Canada turn brilliant red in early fall and leaf out again in April. Those native to the

Plants, like this fall ixia (*Nemastylis floridana*), time their flowers to open only when their primary pollinators are active. The blooms of fall ixia, also known as the "Happy Hour" flower, open for only about three hours in the late afternoon, each for one day.

Everglades rarely get real fall color; they lose their leaves in December and begin putting out new ones within several weeks at most.

Each local population is triggered by its innate photoperiodism, but each is genetically programmed to react a bit differently. When using plants like these in a landscape, special care should be taken to purchase specimens genetically similar to the local population. Specimens originating from disparate locations will never make the adjustment completely and will forever try to respond to changing day-length patterns as if they were living in the place of their origin.

Sometimes we inadvertently disrupt our plants' abilities to use their innate photoperiodism through our landscape practices. As this system of photoperiodism is finely tuned and important to a great many plant responses and behaviors, it is only logical that actions taken to disrupt these cycles might have equally significant reactions. Research at Purdue University found that photoperiod responses in plants are induced under very low light intensity. As might be expected, this is influenced by the type of light bulbs used. For reference, most indoor lighting is ten times more intense than that experienced by plants during a full moon. It is that level of light that plants have evolved to ignore during the evening. More intense light has an influence. A 100-watt incandescent bulb positioned five feet away from a plant provides one-third the light energy of a 150-watt fluorescent cool-white bulb. Landscape lighting may be aesthetically desirable, but it can seriously affect the way your plants perform. Continuous lighting is potentially more damaging than lighting that is turned off sometime during the evening. In addition to its potential effect on growth and flowering, the foliage of trees grown in continuous lighting may be more susceptible to air pollution and water stress during the growing season because the stomatal pores in leaves and green stems remain open.

The spectrum of light emitted by the bulbs is also important. As phytochrome pigments detect light only in the red portion of the spectrum, bulbs that emit low amounts of red light will not stimulate a response even if they emit relatively intense light. Not all lighting is equal. Table 1 provides a general summary of the types of bulbs often used in landscape lighting and their relative impact on plants.

From this, it is apparent that the use of florescent, mercury vapor, and metal halide bulbs has little effect on landscape plants; incandescent and high-pressure sodium bulbs can be quite damaging. In the early days of street lighting, the lamps used most commonly were low-intensity whit-

Table 1. Impacts of light source on landscape plants

Light source	Wavelengths	Potential effect
Florescent	High blue/Low red	Low
Incandescent	High red	High
Mercury vapor	Violet/blue	Low
Metal halide	Green/yellow/orange	Low
High-pressure sodium	High red	High

ish incandescent filaments, higher-intensity bluish fluorescents, mercury vapor, or metal halide lamps. While these sources attracted insects, they had little effect on plants because of the low levels of red light they emitted. In the mid-1960s, high-pressure sodium lamps were developed that emit considerable high-intensity light in the red and infrared regions. Increased injury to woody plants has been reported since the widespread introduction of this type of artificial light. If you must use landscape lighting, choose your bulbs carefully to avoid impacting your plants. Position the lights at least five feet away, point them away from the foliage to the extent possible, and set a timer so they are not on throughout the evening hours.

2

Water

Water is essential to everything in a plant's daily life, and plants have evolved many strategies to acquire sufficient amounts of water and to conserve the water they have once it has been absorbed. As gardeners, we think we understand our plants' needs when it comes to water, but we are often wrong. When a plant is not thriving, we frequently turn to the hose or watering can to solve the problem, but we likely kill as many plants from too much water as from too little. Therefore, if we are to be successful gardeners, it is critical that we understand the relationship plants have with water.

Plant scientists estimate that plants are comprised of 80–85 percent water, a percentage higher than the 40–70 percent estimated for animals. Maintaining a proper water balance is therefore crucial. Plants have unique organelles in each of their cells specifically designed to store water known as vacuoles that shrink and expand depending on how much water is available. When plants have all the water they need, the vacuoles expand and occupy up to 90 percent of the area within each cell. When water is scarce, the vacuole will lose some of its water and shrink, causing the plant to wilt. This expanding and shrinking is a daily phenomenon as plants seek to maintain their water balance in environments that constantly challenge them to do so. Air temperature, relative humidity, and wind speed, as well as the obvious impact of soil moisture, all play roles in varying degrees.

In a normal day, plants take up huge amounts of water simply to maintain themselves. A mature tree, for example, may pull fifty to one hundred gallons of water up through its roots each daylight hour while it is photosynthesizing. Most is lost almost as quickly as it is gained. Experiments have shown that 90–95 percent of the water that plants absorb through their roots is lost as water vapor through the stomata in their leaves and green stems by a process known as transpiration.

It may seem exceedingly wasteful to lose 90–95 percent of the hundreds of gallons of water absorbed by the roots each day, but transpiration has adaptive value. The "misting" this water vapor performs as it escapes through the stomata serves as a sort of air-conditioning that helps cool the plant. Humans sweat, dogs pant, and plants transpire. We are not so different after all. How much cooling is achieved through transpiration depends on a great many factors and has been the subject of some debate among plant scientists, but it is generally agreed that it can lower the temperature by at least five degrees Fahrenheit. Of course, transpiration does much more than keep the plant cool; it is the method by which water, with all its dissolved nutrients, is carried upward from the roots to the stems and leaves. That is its most significant adaptive function.

Each day, plants pull an inordinate amount of water from the ground through their roots and lose it through their open stomata. The 5–10 percent that remains in the plant is not trivial just because it is small relative to what is lost. Retained water facilitates most of the functions necessary for life.

Water is required for photosynthesis. It appears on both sides of the photosynthetic equation that every student of botany is made to memorize. Plants use the water they have inside them (usually pulled into the plant by its roots) to complete the process of photosynthesis. The water is then re-formed and released as vapor once the cycle is over. Sunlight splits carbon dioxide to acquire carbon for glucose production; water is split to acquire the hydrogen also necessary to manufacture it. Because of this, only half of the water used in photosynthesis is used to build more plant tissue. However, all growth and reproduction is dependent on it. No water, no photosynthesis; no photosynthesis, no ability to maintain core metabolic functions—much less grow and reproduce.

Water also gives plants their physical structure. Water-stressed plants wilt as the vacuoles in each cell start to shrink. Scientists call this "losing turgor." I liken vacuoles to water balloons. When water is sufficient, the vacuoles are like a fully expanded water balloon. In this condition, they stretch out each plant cell, and this keeps the leaves and stems erect. Plants try desperately to maintain turgor and avoid wilting whenever possible, but eventually the vacuoles lose water pressure when water becomes scarce and the plant wilts. If the plant is herbaceous (wildflowers, grasses, etc.), the entire plant wilts when water becomes limited. If the plant is woody, wilting is mostly observed in the leaves. Wilted leaves fold downward, and this impairs their ability to receive sunlight for photosynthesis.

Water-stressed plants, like this sweet potato (*Ipomoea batatas*), are easily identified by their wilted leaves and stems.

Wilt can occur even when soils are reasonably moist if water loss from transpiration exceeds the plant's ability to pull water up from the soil with its roots.

Many plants native to tropical and semitropical areas have adapted to this climate by altering their photosynthetic pathways to maximize their efficiency in an environment where midday wilting and carbon dioxide deprivation cause problems to typical C3 plants. These so-called C4 plants initiate photosynthesis just below the leaf and stem surfaces like typical C3 plants do, but they bypass the initial use of the photosynthetic enzyme RuBisCo used by C3 plants in favor of another. RuBisCo evolved over millions of years in an environment where acquiring enough carbon dioxide was never a problem. It is woefully inefficient, however, and prone to using oxygen instead when carbon dioxide levels are low, putting the plant into photorespiration. C4 plants use different enzymes in the chloroplasts just below the leaf surface that are capable of capturing more carbon dioxide from the atmosphere than C3 plants can. They then release this additional carbon dioxide to a second set of chloroplasts deep inside the leaves and stems where they are more immune to the presence of atmospheric oxygen. Here, RuBisCo and the typical C3 pathway operate to produce glucose and release oxygen. It does so more efficiently because it is operating under higher levels of carbon dioxide than it would otherwise, and the risk of photorespiration is greatly reduced. C4 plants can more efficiently fix carbon in drought, high temperatures, and conditions of reduced carbon dioxide than C3 plants, and it seems to be a relatively recent evolutionary adaptation.

Many desert plants take this a step further. Because water is normally scarce regardless of the time of day or year, they keep their stomata closed during the daytime and open them at night when temperatures are coolest and the potential for water loss via transpiration is most reduced. These so-called CAM plants store carbon dioxide while the stomata are open at night and then use it to complete the photosynthetic process during the day, when sunlight is striking the chloroplasts. CAM plants are especially efficient at conserving water during photosynthesis, but it comes with a trade-off in growth rate. There is limited storage space inside the plant's cells for carbon dioxide, and CAM plants cannot expand at night like a balloon. Once the cells are saturated, additional carbon dioxide has no impact. Unlike other plants that access more carbon dioxide as they use it up during photosynthesis, CAM plants have a very limited supply that can't be replenished until the next night after the day's photosynthetic activity is over. CAM plants, such as cactus, are inherently slow growers even in a landscape where they can be given ample amounts of water. It is a result of their genetics that have programmed them to shut their stomata during the daytime regardless of the available water supply.

Soil salinity also affects the ability of plants to take in water. Very few

Cacti have evolved many specializations to deal with extreme drought and temperatures.

plants are truly salt tolerant. As salinity rises, water is pulled out of plant cells into the surrounding environment in an effort to bring everything into balance. Mangroves and other salt-tolerant plants are rare in nature and have evolved complex mechanisms to cope with this tendency to lose water to the outside environment. Some simply close off their ability to absorb salt water when inundated by it; others pump the salts out. Most, however, have an exceedingly difficult time coping in any soil or situation where salts are present. When faced with salts, most plants exhibit stunted growth or die.

Soil salinity affects plants in much the same way as if they were grown in arid conditions because even small amounts of salt in a plant's root zone causes water to leave the plant and creates a water debt that affects photosynthesis and other necessary functions. This can be important in landscapes close to the coast or in areas like Salt Lake City that were once submerged by salt water. It is also a factor for many landscapes that use recycled wastewater for irrigation. Recycled wastewater is typically higher in salinity than fresh water because salts are already present in the water when it arrives to the facility. We think of table salt (sodium chloride) as "salt," but mineral salts all act on plants the same way, and these include calcium, magnesium, potassium, and sulfate salts as well. We add salts to the wastewater stream by our use of water softeners, detergents, soaps, shampoos, liquid fabric softeners, and various other cleaning products. All of this salt-laden water flows through the sewer system to the waste-water treatment plant, and because it is already dissolved when it arrives there, it cannot be effectively removed by the typical current wastewater treatment process. As a result, much of the salt simply passes through the treatment plant and ends up in the treated wastewater that we might irrigate with. Residential and commercial recycled waters vary from under 400 to 900 ppm. Rainwater is not pure but contains only traces of salt under normal situations.

Such increased salinities are usually not enough to affect most plants by themselves, but they tend to accumulate over time within the root zone if present for extended periods, especially in times of low rainfall. These soils need to be rinsed periodically with fresh water/rainwater to counter salt accumulation if sensitive plants are to be maintained with wastewater irrigation.

RESPIRATION

Respiration is essentially the opposite of photosynthesis. Plants take the energy they have stored as simple or complex sugars and burn it in the presence of oxygen to generate the energy they need. Respiration requires oxygen, and it releases carbon dioxide. It also requires water.

Though photosynthesis occurs only under favorable conditions and, in most plants, only during daylight hours, respiration occurs continuously—twenty-four hours a day. Just as with animals, plants have to maintain a baseline metabolism, and to do that, they respire. If they are able to store more energy than that, they can use it to grow and reproduce. Studies by plant biologists have only begun to unravel the effects of water stress on respiration. Most have demonstrated that water stress reduces growth as well as root and flower function, but there are discrepancies. What seems clear, however, is that plants can alter their baseline metabolism to compensate for reduced water availability if changes in soil moisture occur slowly. Extreme water scarcity over the course of just a few days does not give plants a chance to correspondingly slow their metabolism, and this plays havoc with their ability to survive.

The limitations of plant growth, caused by insufficient water availability, are mainly dependent on the balance between photosynthesis and respiration. These processes are intimately linked. Plants genetically altered to reduce their respiration rates correspondingly alter their rates of photosynthesis. Plants also seem to need an increased respiration recovery time to counter a period of decreased photosynthesis due to water stress. Plant growth, therefore, is dependent on the rates of both photosynthesis and respiration, and both are limited by a reduction in soil moisture.

Though maintaining sufficient soil moisture is critical to plant respiration, too much water can also negatively impact plants. Soils are a delicate balance of soil particles and space; the spaces between each particle can be composed of air or water. Both air and water are necessary for soil function, but both can be limiting. Too much air means too little water, and too much water reduces the availability of oxygen for roots to respire. Most plants can adapt to short-term flooding and will recover once the water recedes. Others—especially those adapted to deserts, chaparral, and other excessively well-drained soils—cannot cope with even short-term

Excessive moisture for extended periods can deprive a plant's root system of needed oxygen and kill the roots. "Root rot" caused by overwatering likely kills more potted plants than lack of watering does.

inundation. When roots cannot respire, they die from lack of oxygen and the plant soon perishes.

Epiphytic orchids are one of the best examples of plants with roots adapted to high oxygen levels. As epiphytes, this group of orchids often clings to the branches or trunks of trees and shrubs. Their roots are not in soil but attached to the outside of the bark, where they either absorb the moisture that runs down after a rain or the water vapor found in the surrounding air. Such orchids have a specially evolved outer covering to their roots called vellum that is extremely efficient at absorbing water, but in nature the roots also have evolved to dry out quickly once the water has been absorbed. Orchids do not naturally grow in potting media. When we grow them at home in a pot and place their roots in such material, we are walking a fine line between mimicking the naturally moist environment most originally come from and drowning them in too much moisture. More orchids die from too much water in home cultivation than from lack of it. Their roots, trapped in continuously saturated media in pots, slowly suffocate, die, and then rot away.

This same thing can occur with any plant, except the most water-tolerant species, when grown too long under saturated soil conditions. Plants will attempt to generate new roots for the ones that die, but often their energy reserves become exhausted and generating new roots becomes impossible. Without healthy roots, they wilt at inappropriate times. If we then water them, the few roots they have may counteract the wilt for a short time, but more of the existing root system will suffocate. If the plant were taken out of its pot or its landscape setting, you would find that the root system is virtually nonexistent.

HOW WATER IS TRANSPORTED

Water enters a plant (all plants except bryophytes) through its roots and makes its way to the plant's internal transport system, where it is carried to the rest of the plant's tissues. I will discuss roots in more detail in chapter 5. At this point, it is enough to simply understand them as highly specialized sponges. Plants use no energy to pull water into their roots, and no energy is used to send it into the vascular system for transport. Everything is dependent on the tension created by the open stomata, the loss of water vapor from the surface of its leaves and green stems, and the cohesive properties of water molecules that keep them together in an unbroken chain. Opening the stomata creates suction, like pulling liquid up a straw. As long as the pull of transpiration is maintained, water will move up this straw until it is released into the atmosphere as water vapor. Known by plant biologists as the "Tension-Cohesion Model," it fully explains how water is moved through all vascular plants. In fact, transpiration and the cohesion of water molecules are theoretically sufficient to fully account for the movement of water in any plant less than 500 feet tall—about 150 feet more than the world's tallest documented tree.

Vascular plants have a circulatory system that is not completely unlike that of higher animals. Water is carried in specialized tubules known as xylem; the sugars, hormones, diseases, etc. are carried in other types of tubes known as phloem. Water moves only one way in a plant: it enters the roots and is carried upward until lost through the open stomata of the leaves and green stems. Xylem cells are actually quite simple in design. They are hollow tubes that form unbroken straws throughout the length of the plant. This design also prevents air bubbles from forming and maintains the siphon effect of the straw. In woody plants and most herbaceous flowering plants, the xylem is a ring of tubes just below the bark or epidermis. This makes it very susceptible to damage by such things as weed whackers and lawn mowers. Damage to only a portion of the xylem can be healed over time. The remaining unbroken tubes will still be able to pull water into the canopy while the others attempt to recover. Should the entire circumference be damaged, however, no water will be able to be carried past this wound, and everything above the wound will die. Damage to the xylem is like any other type of injury to an animal. Slight cuts and

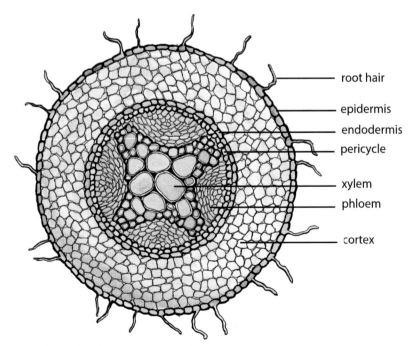

root hair

epidermis
endodermis
pericycle

xylem
phloem

cortex

Typical vascular plant root cross section.

abrasions hardly impact the plant's well-being; acute damage will severely affect the plant's ability to recover; and greater damage will eventually be more than the plant can cope with.

Monocots (grasses, palms, orchids, etc.) do not maintain their xylem in a ring just below the stem's surface. Their xylem tubes are scattered throughout the stem in no discernible pattern. For this reason, palms react quite differently if their outer trunk gets damaged. Though damaged parts may die and slough off, their ability to carry water upward to the crown is much less affected because many of their xylem tubes are deeper inside the stem.

The straw-like cells that comprise the xylem die when fully mature. There is no reason for them to be alive at this point as their function does not require it. That is why cut flowers and Christmas trees can still pull water out of a vase or tree stand even though the plant has lost its roots and much of the rest of its support structure. Some plants can generate new roots after their stems are cut and defy death. Their ability to do this depends on the genetics of the plant, where the cut has been made, and the age of the cut part (i.e., new versus mature growth). Most plant cuttings can bide their time for a while without a root system, but eventually they will die without one. They need to add new xylem as they grow and be able to repair damages to existing xylem cells, but these cells are not made by the xylem itself. Once formed, xylem cells die, and they function to transport water in this state.

DISSOLVED NUTRIENTS

The water carried in a plant's xylem is far from pure; it contains many dozens of dissolved elements necessary for plant life. Nutrients occur in the soil, but they must be dissolved in water to be carried inside the plant. Of the more than ninety naturally occurring elements present in soil, a little more than sixty are routinely found in plant tissue. Most provide no biological purpose, but nineteen are considered essential. Plants cannot survive without them (table 2). Though plants feed on sunlight, these nineteen essential elements are the "multivitamins" necessary to support life functions. Each supports specific functions and is absolutely necessary in some amount.

Fifteen essential nutrients come directly from soil and must be dissolved in water to be available; the remaining four (nitrogen, oxygen, hydrogen, and carbon) originate in the atmosphere or as water and get fixed into plant tissue in a variety of other ways. The importance of soil will be discussed in more detail in chapter 3, but suffice it to say that not all soil is created equal, and different parts of the earth have different percentages of these fifteen elements present. It is this fact that makes soil fertility so variable, not the amount of sand or organic matter, for example.

Water dissolves a small portion of these essential elements as it passes through the soil as rainwater or melting snow. It then has to be captured by the plant's root system before it leaches beyond its reach. It is a constant battle; gravity pulls water downward while small electrical charges on the surface of each soil particle hold the water molecule and prevent gravity from taking it away. The smaller the grain of soil, the stronger the electrical charge. Clays have a much stronger hold on water droplets with their dissolved essential elements than sands. This makes the leaching rates of essential elements much slower in clay soils than sands, but it also makes it more difficult for plants to pull the water off their surface.

The availability of essential elements, therefore, is predicated on a complex suite of factors. Assuming they are present to begin with, too much rainfall can wash them away temporarily, and extremely small soil particle sizes can hold them too tightly to be available. The solubility of these elements is also a factor of the soil's pH. The pH scale is one that measures how acidic or basic something is from a scale of 1 to 14. Neutral is measured at 7; acidic soils are less than 7, and basic soils are greater. Most soils range from 6 to 8. As soils become more acidic, they make certain elements more likely to dissolve. Iron is a good example. Iron in the parent soil material becomes increasingly available as the pH decreases below 7. So does aluminum and manganese. While both iron and manganese are essential elements, too much of a good thing can become toxic. Some plants are especially well adapted to acidic soils and tolerate these higher iron, manganese, and aluminum levels. Members of the Ericaceae family are especially well-known examples. Ericaceous plants include blueberries (*Vaccinium* spp.) and azaleas (*Rhododendron* spp.). Such species often decline in high pH soils because they don't get the levels of iron and manganese that they require and special fertilizers for them are designed to rectify this.

Table 2. Essential plant elements

Element	Function
MACRO-NUTRIENTS	
Carbon	Lipids, carbohydrates, nucleic acids, proteins
Hydrogen	Lipids, carbohydrates, nucleic acids, proteins
Oxygen	Lipids, carbohydrates, nucleic acids, proteins
Nitrogen	Significant part of proteins, nucleic acids, and chlorophyll
Potassium	Important to the function of cell membranes, including turgidity and stomatal opening and closing
Phosphorus	Component of nucleic acids, phospholipids, and short-term energy storage (ATP/ADP)
Calcium	Important in plant physiology, plasma membrane permeability, and building the "cement" that holds adjacent cell walls together
Magnesium	Significant part of the chlorophyll molecule
Sulfur	Important to structure of many amino acids and vitamins
Silicon	Enhances growth and fertility, reinforces cell walls
MICRO-NUTRIENTS	
Chlorine	Maintenance of cell turgor
Sodium	Maintenance of cell turgor
Iron	Associated with various enzymes and catalysts in photosynthesis and normal cell function
Manganese	Associated with various enzymes and catalysts in photosynthesis and normal cell function
Zinc	Associated with various enzymes and catalysts in photosynthesis and normal cell function
Copper	Associated with various enzymes and catalysts in photosynthesis and normal cell function
Nickel	Associated with various enzymes and catalysts in photosynthesis and normal cell function
Molybdenum	Associated with various enzymes and catalysts in photosynthesis and normal cell function
Boron	Present in cell walls, involved in nucleic acid metabolism and cell growth

In high pH soils, these essential elements are much less soluble, but many plants can still get all they require. Other elements, however, often prove limiting. Organic "salts" such as calcium phosphate do not adequately dissolve when the pH gets too basic. Most plants require ample amounts of both calcium and phosphorous. It takes a specially adapted one to tolerate the reduced levels found in high-pH soils. Here in my adopted state of Florida, our soils lie on a foundation of limestone–calcium carbonate. Pine needles and other leaf litter, coupled with the slightly acidic nature of rainwater, make the overlying soil acidic. As this acidic water reaches the underlying limestone, it dissolves it, and the layer of soil immediately above it becomes quite basic. Areas of the state where limestone is near the surface produce basic soils; areas of deep soil are acidic. Both situations produce their own plant communities, and only a few plant species are adapted to all extremes.

3

Soil

Soil is a complex association of organic and inorganic materials, water, air, and a huge variety of living organisms. In the simplest terms, it is the medium from which plants grow, the thing that provides them anchorage, the stuff you dig a hole in at planting time, or the bag of material you purchase to put in an empty pot. The reality, however, is that soil is much more than this, and understanding more of its intricacies will make you a better gardener.

SOIL COMPOSITION

In most plant biology textbooks, soil is defined as the "dead" portion of the overall soil equation. It is a composition of inorganic and organic soil particles and the sum of the space comprised of air and water. In most soils, the particles and the space each comprise about half of the total. The ratio of water to air depends on how saturated the soil is at any one moment, and it shifts from one extreme to the other based on rainfall and the depth of the water table. The ratio of organic to inorganic particles varies little in most soils within the rhizosphere (the area/depth of the soil where roots are found), though organic matter often accumulates in saturated soils near the surface because decomposition is hindered in these conditions.

The inorganic portion comprises about 45 percent of most soils. It is derived from the erosion of parent rock, laid down at the time of the earth's formation more than 4 billion years ago. As land plants did not appear until 2.5 billion years later, soil had plenty of time to develop and accumulate before it needed to be used. The earth's formation did not leave the basic elements uniformly scattered across its crust, and different levels of volcanic

Typical soil in my adopted home state of Florida is a combination of sands, silts, clays, and organic particles. Most evident are the grains of sand.

activity, erosion, and tectonic uplifting have made the distribution of these elements even more variable. Changes in weather patterns and sea level altered this pattern even more. Over the millennia, the basic composition of parent material has been altered dramatically from one area of the earth to another, often from one neighborhood to another. The earth's surface is not a level playing field when it comes to plant nutrients.

Geologic and weather activity, followed by the activity of fungi, lichens, and plants, breaks solid rock into smaller and smaller particles. Soil scientists consider any particle larger than 2 mm (about $\frac{1}{16}$ inch) to be "stones" and functionally useless to plants in terms of providing nutrients. Stones affect drainage and a variety of other functions, but they are too large to interact with plants on a nutrient level. The portions that are considered soil are the particle sizes 2 mm or less in diameter. Soil scientists break these down into three basic size classes: sand (2–0.02 mm); silt (0.02–0.002 mm); and clay (< 0.002 mm). Particles whose actual grain can be seen by the naked eye are sand. Silt is like ground flour; you can see flour, but grains are too small to see individually. Clays are particles so small that their individual grains cannot be seen without an electron microscope. When I put clay between my fingers, wet it, and then rub it, it forms a smooth ribbon that exhibits no sign of grittiness whatsoever. The individual grains of clay are too small to feel.

Soil particle size, therefore, has very little relationship to soil fertility. The fertility of the soil is dependent only on the presence of the fifteen essential elements derived from the parent rock; the availability of nitrogen, carbon, oxygen, and hydrogen; and their accessibility to the plant. Sand is not inherently infertile. Florida's sandy soils, for example, support one of the richest diversities of native plant species in the nation—far more diverse than the so-called fertile prairie soils of the Midwest. Essential elements may leach faster in Florida sands than in the loamy midwestern prairie soils, but they are not inherently infertile. Natural fertility is merely a function of parent material and its accessibility. Accessibility is mostly dependent on soil chemistry and the leaching rate.

The influence of soil pH on the availability of certain essential elements was discussed above. Leaching rates are a function of soil particle size and weather. To be precise, leaching is the movement of water through the soil column. Water on the soil surface moves to increasingly greater depths until it reaches depths beyond the reach of plant roots and eventually into an aquifer, where it is stored long term. As the water moves through the soil, it carries dissolved elements such as iron and potassium with it. It also carries organic matter and the tiniest clay particles. In certain soils where the water table is not exceptionally deep, these elements and particles are deposited in discernible layers several feet beneath the surface. In exceptionally deep, well-drained soils, they are often carried so far below the surface that their deposition cannot be detected.

In most native soils, elements are constantly being added by their dissolution from the soil particles themselves and by being made accessible again from the decay of plant and animal material near the surface. As they are added, however, they are leached deeper and deeper into the soil column. Most leached elements get absorbed by plant roots and are then sequestered inside the living plant until it dies and decomposes. This rough balance in soil fertility is often upset in developed landscapes, especially when plants are harvested and not allowed to decompose on-site. Modern agriculture is perhaps the worst culprit in this scenario. We often call such soils "overworked," but it is not the "work" that upsets the balance but the fact that the plants are routinely removed with all their sequestered essential elements and taken out of the system. Over time, the parent material cannot replace the amount lost. These elements then need to be returned by adding fertilizers.

Urban soils also lose their native fertility because of our traditional landscaping practices. In most parts of the industrialized world, we remove valuable dead plant material from the landscape, bag it, and throw it away in places where it will not fertilize soil of any kind. Lawns are the single-greatest source of this type of loss. Mowing lawns and discarding the clippings off-site removes millions of tons of valuable organic material from urban landscapes annually. This is further exacerbated if we've also added commercial fertilizers to make the grass more luxuriant. We're throwing this fertility away too. Raking landscapes of leaves and debris does the same thing. There is no reason to remove this valuable fertilizer and cast it away.

Soil fertility is replaced each season by the decay of organic material such as leaves and pine needles. Removing this nutrient source deprives soil of a significant source of fertilizer.

If it can't be kept on-site, it should be composted and then returned. We also need to rethink our use of mulches within our planting beds. Mulches, comprised of nonorganic or woody materials, do not decay easily or replace the natural fertility found in dead leaves. Each leaf and blade of cut grass is essentially a packet of fertilizer, neatly wrapped and designed to replace what the original plant removed from the soil. We need to quit thinking of this as lawn "waste." It is anything but that.

Although particle size has very little influence on soil fertility, it has a major impact on other aspects of plant growth. For one, it impacts drainage, affecting the delicate balance of oxygen and water within the root zone. Water moves through sandy soils more quickly than through soils high in clays, therefore the essential elements leach more quickly too. Soils high in clays retain these elements much longer. The microscopic soil particles hold the water droplets tighter to their surface. The essential elements remain in the root zone longer before they leach too deeply to be used. The downside is the lack of oxygen available in this root zone once the soil is fully saturated. In general, plants adapted to sandy soils have to be extremely efficient in gathering essential elements before they are leached too deeply, and they have roots that need high amounts of oxygen for respiration. Plants adapted to clay soils are on the opposite end of the spectrum. Some plants, like butterfly milkweed (*Asclepias tuberosa*), occur in all ranges of soil types within their natural distribution, but the plants in each geographic region are very different in terms of their soil needs. The butterfly milkweed of my Wisconsin youth occurs in deep, rich prairie soil; the same plant in Georgia occurs along roadsides in red clay; and in Florida it is found only in well-drained sands. Although technically the same species, you cannot grow it in a soil type it did not evolve with. I've tried, and I've failed every time.

Particle size also affects root growth. Though roots will be discussed much more thoroughly in chapter 5, the relationship between root growth and particle size needs to be mentioned here. As roots grow, they push forward, seeking paths of least resistance. Sandy soils, with their many air spaces, allow roots to grow largely unimpeded. Soils high in clays do the opposite. As clays swell and dry, they often form a mostly impenetrable barrier to root growth. Dry clay layers sometimes crack and allow roots to chisel their way through, but at other times they stop deeper penetration into the soil column. Most plants grown in high-clay soils develop very shallow root systems that spread a great distance laterally from the main stem.

Besides the particles originally derived from the parent rock, soil contains some fraction of decomposed organic material. In what is considered "typical" soil, the percentage of organic material is about 5 percent of the total composition. Organics are derived from decayed plant and animal material. They are unavailable to plants as a supply of essential elements until they've been broken down by an assortment of soil organisms—everything from bacteria and fungi to termites, worms, and burrowing animals.

Organic material influences soil in a variety of ways, but it is not necessarily fertile. Gardeners often look at organic-rich soils and infer fertility, but the elements they provide plants are completely dependent on the composition originally sequestered by the decayed plant or animal. Essential plant elements don't simply materialize out of decaying plant parts or decomposing dung. They had to be there to begin with. Plants grown in nutrient-poor conditions will have fewer nutrients to impart when they die than plants grown in fertile soils. Leaves shed in the late fall will have fewer nutrients than green leaves removed during the peak of growth. Wood will have fewer nutrients than herbaceous material. Compost adds organics to the soil, but the amount of essential nutrients it adds is largely unpredictable.

Organic matter derived from plants holds water like a sponge once wetted, but it repels water when dry. This can have a profound influence on soil moisture and root growth. In already-moist soil, organics soak up moisture and hold it tightly. They remain wet a lot longer than most inorganic soil particles, and they can help plants make it through short periods of drought better than if they had not been present. Once dried, however, organic particles actually repel water until they are rewetted.

This is one of the big dilemmas for gardeners to consider when adding new plants to the landscape. Typical commercial soil mixes are largely composed of peat and other organics. Such mixes hold water well and help prevent the plant from drying out too quickly and requiring more frequent watering. In the nursery where they are grown, this works out well. When these plants are transferred into the ground, however, problems can ensue if the root ball is allowed to dry out or the region surrounding the root ball is not kept sufficiently moist. Container-grown plants develop a

Organic particles in soil are evident as dark flecks among the mineral components. These particles significantly affect how water moves through the soil.

well-defined root ball inside their pot. For those roots to leave that media, the soil around them has to entice them out of it. If it is too dry, they will not venture into it. If the root ball itself is allowed to dry out around the edges, the new root growth at the edges of the root ball will die and there will be no live roots there to venture out with. New roots will not grow downward into their new planting site if the soil below the root ball is dry either. Organics provide benefits to the nurseryperson, but they can complicate life for the gardener. Gardeners adding new plants to their landscapes must remain vigilant in keeping both the root ball and the soil around it evenly moist until the plant is thoroughly acclimated to its new surroundings.

Gardeners often think to supplement their soil at the time of planting by adding organics to the hole. More often than not, this complicates the ability of plants to adjust to their new home; at the extreme it can eventually kill them. Soggy organic soil suffocates new roots while dry organic material repels water and retards establishment. Natural soils are rarely comprised of more than 5–10 percent organic particles. More than that is more than most root systems can naturally deal with.

Realize also that the organics you start with will eventually decay and disappear. The inorganic matter will remain, possibly surrounded by large air spaces that were once occupied by the now-decayed organic material. In most situations, it is best to plant directly into your native soil. Some recent research even suggests that the potting soil surrounding the root ball should be washed off at the time of planting to assist with establishment. Of course, this would have to be done carefully. The fine feeder roots are easily damaged by rough handling or aggressive water action.

SOIL ORGANISMS

The textbooks I use to teach plant biology restrict the definition of soil to include only the nonliving components; their discussion stops after consideration of the inorganic and organic particles, and the water- and the air-filled spaces. When thinking of soil, it is imperative to also include the living organisms that reside in it. They are critical to how plants function in soil. They provide vital services that allow plants to function at their best, but they also pose risks to plants' overall success. Their presence gives soil its life; their absence creates a largely sterile wasteland that makes plant life extremely difficult. How we manage our soil makes a great difference in how our plants will function. We mismanage the living component at very great risk to the long-term health of our plants.

It has been argued that the single-greatest leverage point for maintaining a sustainable and healthy future for the world's human population lies in the living component of our soil. These living organisms include a vast array of microscopic and macroscopic things. We are coming to fully comprehend the role that our gut biota play in human health and the concept of a diet based on probiotics. A similar understanding is growing of the importance of soil microorganisms in the health of soil and the plants it sustains. Over the past century, we have unwittingly wreaked havoc with our use of certain fertilizers, fungicides, and herbicides—especially in modern agriculture and urban landscapes. Healthy plants require living soil at least as much as they require the nonliving components.

Bacteria

Healthy soil contains a varied suite of soil bacteria. Bacteria are tiny, one-celled organisms, generally ⁴/₁₀₀,₀₀₀ of an inch wide (1 μm) and somewhat greater in length. What bacteria lack in size, they make up in numbers. A teaspoon of productive soil generally contains between 100 million and 1 billion bacteria. A ton of microscopic bacteria may be active in each acre of soil. Bacteria are tough. They occur everywhere on earth and have even been found more than a mile beneath the earth's crust.

Bacteria can be separated into five major groups. Most are decomposers that consume simple carbon compounds such as root exudates and fresh plant litter. Through this process, decomposer bacteria convert the energy contained in soil organic matter to forms useful to plants and other living soil organisms. Some decomposer bacteria can break down pesticides and pollutants in soil. Their presence keeps soil healthy. Decomposer bacteria are also especially important in keeping essential nutrients within the root zone. As bacteria use many of the same nutrients needed by plants, they sequester large amounts that might otherwise leach deeper into the soil column.

A second group of bacteria are the ones that form partnerships with plants. The most well-known of these are the free-living nitrogen-fixing bacteria known as cyanobacteria or blue-green algae and the symbiotic ones that most famously include the *Rhizobium* spp. that form nodules on the roots of legumes. Cyanobacteria are especially important to soil fertility, and they are arguably the most successful group of microorganisms on earth. They are the most genetically diverse. They occupy a broad range of habitats across all latitudes, and they are even found in the most extreme places on earth, including hot springs and hypersaline bays. Cyanobacteria preceded most other life on earth, and they created the conditions in the planet's early atmosphere that directed the evolution of aerobic metabolism and photosynthesis in algae and plants. As photosynthetic organisms, they create carbon-based sugars from sunlight in the presence of water and carbon dioxide. As they die, these nutrients are made available to higher plants.

Their major contribution to soil, however, is their ability to convert atmospheric nitrogen into nitrogenous compounds useful to plants (and ultimately, to animals). As much as 80 percent of the earth's atmosphere is composed of nitrogen gas (N2). We are awash in it, but it is inert in this form and cannot be assimilated into plant or animal tissues. This is a huge problem because so many of our tissues, enzymes, and other life-sustaining functions require nitrogen as a building block. All the biologically useful nitrogen on earth originates from the types of bacteria that capture inert nitrogen gas from the air and convert it to ammonia and nitrogenous salts.

The third group of bacteria is comprised of pathogens. Bacterial pathogens include several genera of soilborne species and some that produce stem and leaf galls in plants. Pathogenic bacteria are found throughout healthy soil. Some are very species specific and interact only with their host species or closely related species. In spite of how common these bacteria are, most plants do not display serious symptoms of disease. Disease usually occurs when conditions have been altered from the norm, or when a soil organism is accidentally introduced where a highly susceptible plant species occurs. Intensive production in agriculture, horticulture, or forestry increases the opportunities for diseases to develop. Planting large stands of the same plant species together also increases the probability of disease outbreak. In many situations and in most healthy soils, plants have a resistance to most bacterial pathogens.

A fourth group, called lithotrophs or chemoautotrophs, obtains its energy from compounds of nitrogen, sulfur, iron, or hydrogen instead of from carbon compounds. Some of these species are important to nitrogen cycling and the degradation of pollutants.

The last group of bacteria is composed of free-living organisms capable of photosynthesis. There are three major types, two known as purple bacteria and one that is green. These types of bacteria are anoxygenic, meaning that they use light energy and convert it to biologically useful products without generating oxygen as a by-product. Most free-living photosynthetic bacteria live in the water, and none are especially important in soil.

Bacteria from all five groups perform important functions related to water dynamics, nutrient cycling, and disease suppression. Some bacteria affect water movement by producing substances that help bind soil particles into small aggregates (those with diameters of $1/10,000$–$1/100$ of an inch or 2–200 μm). Stable aggregates improve water infiltration and the soil's water-holding ability. In a soil with a diverse bacterial community, most

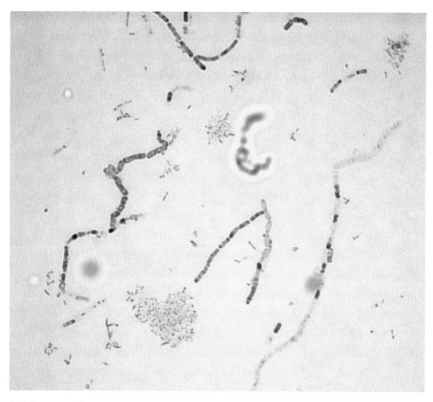

Soil bacteria play a very important role in the relationship between plants and soil. Photograph by Mira Janjus.

soil functions are vastly improved, and the likelihood of soilborne diseases is significantly reduced.

Different species of bacteria thrive on different food sources and in different microenvironments. In general, bacteria are more competitive when located near easy-to-metabolize foods. This includes fresh, young plant residue and the compounds found near living roots. Bacteria are especially concentrated in the rhizosphere, the narrow region next to and in the root. There is evidence that plants produce certain types of root exudates to encourage the growth of protective bacteria near them.

Bacteria alter the soil environment sufficiently to favor certain plant communities over others. Before plants can become established on fresh sediments, the bacterial community must establish first, starting with photosynthetic bacteria. These fix atmospheric nitrogen and carbon, produce organic matter, and immobilize enough nitrogen and other nutrients to initiate nitrogen cycling processes in the young soil. Then, early successional plant species can grow. As the plant community is established, different types of organic matter enter the soil and change the type of food available to bacteria. In turn, the altered bacterial community changes soil structure and the environment for plants. Some researchers think it may be possible to control the plant species in a place by managing the soil bacteria community.

Certain strains of soil bacteria have antifungal properties that inhibit some plant pathogens. These bacteria can increase plant growth in several ways. They may produce a compound that inhibits the growth of pathogens or reduces the likelihood of infection. They may also produce compounds (growth factors) that directly increase plant growth. These plant-growth-enhancing bacteria occur naturally in soils but not always in high enough numbers to have a dramatic effect. In the future, it may be that agriculture and horticulture will supplement the soil with additional soil bacteria of this type.

Fungi

Healthy soil is also full of fungi. It is estimated that one gram (about 0.04 ounces) of soil contains approximately 1 million fungi, everything from single-celled yeasts to multicellular structures. Though we often think of fungi as organisms that cause rot and disease, healthy plants require them in the soil. They are every bit as important in soil as bacteria.

Fungi have no chlorophyll and are not able to photosynthesize. They cannot use atmospheric carbon dioxide as a source of carbon and therefore are like animals and require organic substrates to get carbon for growth and development. They get that by spreading out in all directions by means of a spider-web-like network called hyphae. These are essentially stomachs that produce compounds that allow them to digest all kinds of organic matter. Different types of fungi have evolved to feed on different substrates. Some can even digest petroleum products and render them nontoxic. A few years ago, scientists working in the Amazon found a species capable of consuming plastics. Fungi are phenomenal organisms.

Soil fungi, especially mycorrhizal fungi, are especially significant in soil.

There are three major types of soil fungi: the decomposers, the symbionts, and the pathogens. Decomposers, or saprophytic fungi, feed on dead organic matter such as leaves and twigs and use it to grow. As they do this, they produce carbon dioxide and organic acids as waste products. By consuming the nutrients in the organic matter, they play an important role in keeping them within the root zone of plants. Without fungi in the soil, dense woody material, such as logs, would accumulate on the soil surface and disrupt the soil ecology of forested areas.

The symbiotic fungi may be the most important to plants, however. These fungi develop mutually beneficial relationships with plants. Known as mycorrhizal fungi, they colonize plant roots and vastly improve their ability to absorb water and nutrients from the soil. Mycorrhizal fungi are the capillary beds of a plant's water-gathering circulatory system. Even the tiniest root hairs are stubby and inefficient in comparison to the hyphae of mycorrhizal fungi.

Water molecules adhere to soil particles due to their electrical charge and the surface tension of water. As soil dries, it becomes increasingly difficult to pull the last remaining water molecules off the soil particles. Roots that have been colonized by mycorrhizal fungi are significantly better at removing this water than they would be otherwise. The potential for wilting is diminished, and plant growth can proceed along with the collection of essential plant nutrients.

There are several major groups of mycorrhizal fungi, but all are common in healthy soil. The most common group is the "arbuscular" types—the glomeromycetes. Arbuscular mycorrhizal fungi actually penetrate the outside epidermal layer of the roots and form an unbroken network with the root's vascular system. Ectomycorrhizal fungi form a net over the surface of the plant's roots. This type is not as efficient for the plant as the arbuscular type, but it still vastly improves the overall efficiency of the roots. Nearly every plant forms relationships with mycorrhizal fungi. Recent research has shown that the developing root network even produces chemicals to attract them. It is a survival mechanism that has evolved over millions of years. This will be discussed further in chapter 5.

Some soil fungi are pathogens, and many are harmful or even lethal to plants. This group includes fungi such as *Verticillium*, *Phytophthora*, *Rhizoctonia*, and *Pythium*. These organisms penetrate the plant and decompose the living tissue, weakening it and making it nutrient deficient. Pathogenic fungi are often the dominant fungi in the soil, but research has shown that soils with a high diversity of other fungi and bacteria naturally suppress soilborne fungal diseases. Healthy root systems inoculated with mycorrhizal fungi, for example, develop the ability to keep pathogenic fungi away or kill them outright when they get in close proximity to the root. Healthy soil produces high soil biodiversity, and this in turn produces roots better able to fight off root pathogens. Soil fungicides can exacerbate a fungal problem in much the same way insecticides can disrupt an ecosystem. Killing the harmful *and* beneficial organisms indiscriminately at the same time provides short-term relief, but the harmful ones return more quickly than the others, and they can multiply unchecked. Healthy soils are rarely lethal to plant roots because of fungi. Fungal disease in the roots is most often a sign that something else is not right with the soil.

Soil Fauna

A multitude of animals also live in the soil, and the majority play important roles in terms of plant health. These include a wide variety of invertebrates (animals without backbones) as well as fossorial vertebrates (animals with backbones that live underground). Of these, the invertebrates are the most important and certainly the most widespread. These include roundworms, segmented worms, and insects.

Roundworms, also known as nematodes, are nonsegmented worms typically $1/500$ of an inch in diameter and $1/20$ of an inch (1 mm) in length. Nematodes are extremely common and are found everywhere. It has been said that if you could take away all the soil and living tissue in the world, everything left would be clearly outlined in nematodes. They occur in every part of our bodies, throughout the bodies of other animals and plants, and throughout the upper reaches of soil occupied by plant roots. Because of their size, nematodes tend to be more common in coarser-textured soils. Nematodes move through the soil in water films that occur between the spaces of soil particles.

An incredible variety of nematodes function at each level of the soil food web. Of these, only a small number are directly harmful to plants; the majority are beneficial. Free-living nematodes can be divided into four broad groups based on their diet: bacterial feeders consume bacteria; fungal feeders puncture the cell wall of fungi and suck out the internal contents; and predatory nematodes eat all types of other nematodes. The other group is classified as protozoa. Protozoa are predators that feed on a wider variety of other small animals. They either eat smaller organisms whole or attach themselves to the cuticle of larger nematodes, scraping it away until the prey's internal body parts can be extracted.

Heavily worked and treated soils tend to have fewer nematodes than native soils and far less species diversity. Grasslands may contain up to five hundred nematodes in each teaspoon (dry gram) of soil, and forest soils generally hold several hundred. Agricultural soils generally support fewer than one hundred nematodes per teaspoon. Native soils commonly contain more predatory nematodes and a higher proportion of bacterial- and fungal-feeding species, suggesting that these beneficial types are highly sensitive to soil disturbances.

Nematodes are important in making many essential nutrients available to

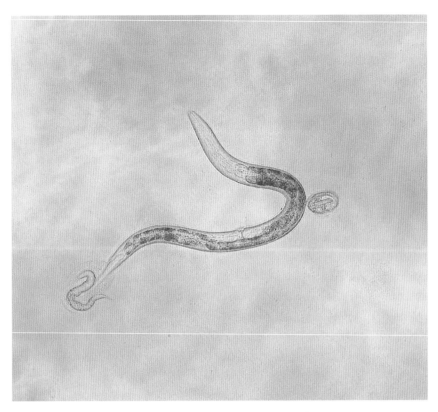

Nematodes in the soil can play positive and negative roles in the plant-soil relationship. Photograph by Mira Janjus.

plants. When nematodes eat bacteria or fungi, there is an excess of nitrogen-based compounds made available because bacteria and fungi contain much more nitrogen than the nematodes require.

Nematodes help distribute bacteria and fungi throughout the soil and along roots by carrying live and dormant microbes on their surfaces and in their digestive systems. They are also food for higher-level predators, including predatory nematodes, soil microarthropods, and soil insects. Bacteria and fungi sometimes parasitize them.

High nematode densities in the soil can also negatively affect plants when they prey on beneficial organisms. Plant feeders suppress plant growth directly; fungal- and bacterial-feeders can reduce the populations of beneficial bacteria and fungi, and this can reduce plant productivity. Some nematodes cause disease. Pathogenic nematodes feed on plant roots or stems and are not, by definition, free living. Those few species responsible for plant diseases have received a lot of attention for good reason. They can cause significant economic loss, especially in agriculture and horticultural settings. Root-feeding nematodes puncture the cell wall of plant root cells with their specialized mouth parts and siphon off the internal contents of the plant cell. Ultimately, the roots are destroyed and the plant dies. Nematode damage to the plant's root system also provides an opportunity for other plant pathogens to invade the root and thus further weaken the plant. Direct damage to plant tissues by shoot-feeding nematodes includes reduced vigor, distortion of plant parts, and death of infected tissues depending upon the nematode species involved.

Predatory nematodes may regulate populations of harmful nematodes, thus reducing their damage. It is a complex interactive world beneath the soil surface, and it is quite easy for us to upset that balance when we disrupt the natural soil processes in managed landscapes.

Nematode control has typically involved the use of chemicals that kill them outright, but such methods disrupt the overall structure of the microfaunal community and indiscriminately kill the beneficial organisms as well as the offending ones. Biological and cultural controls are preferable, especially in landscape and natural settings. The most practical form of biological control is the use of nematode-resistant plants, especially plants native to your geographic region. Native plants have evolved to live in balance with native soil fauna. Their associated mycorrhizal fungi actually produce compounds to deter nematodes from attacking them. This is not often the case with non-native plants. They have not had time to evolve such complex relationships with local nematode and mycorrhizal species.

Effective cultural controls are most effective in agricultural settings and involve using crop rotation to disrupt the proliferation of pathogenic nematodes in the soil. Planting nematode-resistant crops for several years following a crop of more susceptible plants can be effective in reducing nematode numbers, but it won't eliminate them completely.

Segmented worms, earthworms, and their relatives are primarily valuable in nutrient recycling. They sometimes are referred to as "ecosystem engineers." Much like human engineers, earthworms change the structure of their environments. Different types of earthworms can make both horizontal and vertical burrows, some of which can be very deep into the soil column. These burrows create pores through which oxygen and water can enter and carbon dioxide can leave. Their droppings are also very important in changing soil structure.

Earthworms play an important role in breaking down dead organic matter in the soil. They do this by eating the organic matter, passing it through their digestive tracts, and releasing it once it has been digested. Bacteria and fungi break down this organic matter even further inside the worm's intestine. By the time their feces leave their gut, a significant amount of nutrients is released into the soil.

Earthworms are also responsible for mixing soil layers and incorporating organic matter into the soil. Charles Darwin referred to earthworms as "nature's ploughs" because of their mixing of soil and organic matter. This improves the fertility of the soil by allowing the organic matter to be dispersed throughout. The organic matter is then made available to soil bacteria, fungi, and the plants themselves. Studies have shown that the presence of earthworms increases the numbers of beneficial bacteria and fungi in the soil. Therefore, the role earthworms play in healthy soil cannot be overstated.

The other major soil invertebrates are insects, namely ants and termites. Though often given a bad image in popular culture, they are indispensable to healthy soil. Ants are also ecosystem engineers, greatly affecting physical, chemical, and biological properties of the soil. The effects on physical soil properties are related to the building of tunnels, which increases soil porosity and may cause separation of soil particles according to their size.

Ant-mediated chemical changes of soil are represented mainly by a shift of pH toward neutral and an increase in nutrient content (mostly nitrogen and phosphorus) in ant-nest-affected soil. These effects result from the accumulation of food in their nests and the effect it has on biological processes such as accelerating the decomposition rates of soil organic matter. Effects on the soil vary between ant species, and substantial variation can be found in the same species living in different climate and soil conditions.

Termites are the subject of countless television commercials due to the very real damage they can cause to homes and other wooden structures. Termite control is a major industry, and most of the public recognizes only their negative side. Truth be told, termites have a positive side as well, one related to their role in the decomposition of woody debris and its impact on soil.

Earthworms are especially important in mixing soil and in processing organic matter into something usable by plants.

Like ants, termites are soil engineers. Not every species of termite is silently waiting to eat your wood-frame home. Some build tunnels and mounds, and this influences soil structure and water infiltration rates. Most termites also increase carbon and nutrient levels in the soil, especially nitrogen, phosphorus, and potassium, as well as exchangeable magnesium and calcium.

All termites are detritivores. They feed on dead woody plants as well as dead parts of living trees and shrubs. A termite's mouth is uniquely designed to tear apart pieces of woody material. This feature makes them quite efficient at breaking down logs and other woody detritus in a forested setting.

Termites play a significant role in breaking down woody debris and making those nutrients available to plant roots.

By quickly recycling large woody debris into easily accessible nutrients, they play a significant role in soil ecology. If you live in a wood-frame home where termites are common, take precautions to keep them away and restricted to your landscape. Keep things like firewood stacked far away from your home and up off the ground, and keep moisture away from the foundation and roof. If you have woody debris in your landscape, well away from your home, let the termites recycle it.

Soil is also modified by larger vertebrates such as burrowing rodents and tortoises. By turning over the soil and increasing its porosity, such vertebrates can have a major influence on soil structure. Most, however, are herbivores (plant eaters), and the damage they do in a landscape setting can outweigh their benefits. Pocket gophers, ground squirrels, and woodchucks eat plants and their root systems, whereas moles and armadillos are invertebrate predators and cause damage only by exposing root systems to air and drying them out. For most gardeners, their presence involves far more problems than benefits.

4

Basic Plant Structure and Growth

So far, we have focused primarily on how the environment affects plants, the so-called abiotic factors. Light, water, and soil play significant roles in how plants work and understanding them will make us better gardeners, but we also need to know how plants themselves function, how they are designed, and how they interact with one another and the world around them.

PLANT CELLS

Like animals, plants are composed of different kinds of cells. These are arranged into tissues, and the tissues are organized into organs. Plants simply have different kinds of cells, tissues, and organs than animals. They have evolved with different needs.

Plant and animal cells are structured very similarly. They both contain membrane-bound organelles such as a nucleus to direct cellular functions; mitochondria to generate energy from stored proteins, carbohydrates, and fats; endoplasmic reticulum to produce proteins; Golgi apparatuses to move material between cells; lysosomes to break down all types of complex biological molecules no longer needed by the cell; and a cell membrane to surround them all. Every one of these functions is vital to living organisms that have evolved beyond the bacteria and virus stages, so it is no wonder that plants and animals share them.

The few differences that do occur, however, are quite significant and reflect the very great differences between the lives of plants and animals. One of the most obvious differences is that plant cells are surrounded by a cell wall that is produced by the inner cell membrane. The wall gives each cell the additional structure plants require. Plant cell walls are composed of cel-

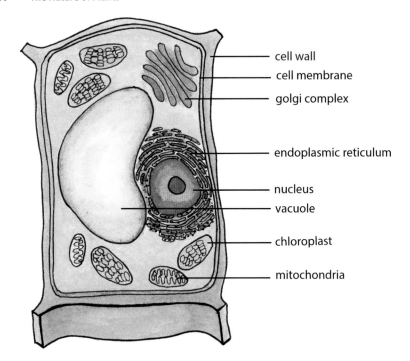

cell wall
cell membrane
golgi complex

endoplasmic reticulum

nucleus
vacuole

chloroplast

mitochondria

Typical plant cell.

lulose, a complex carbohydrate built from the glucose molecules produced in photosynthesis. Cellulose is an ideal building block for plants; it is tough but flexible, and it absorbs water. The next time you use a tissue or paper towel, be thankful its foundation is cellulose instead of the chitin that forms the cell walls of fungi.

Plant cell walls are further strengthened by strands of pectin that are woven throughout and then mortared together with the adjacent cell wall with a layer of additional pectin called the middle lamella. This arrangement provides both the strength and flexibility plants need, but it restricts the ability of important materials to move from cell to cell. Plants solve this dilemma by having tiny pores called plasmodesmata between adjacent cells. Thus, plants are structured like a brick building. The cells are stacked one atop another, mortared together by layers of pectin, but are able to communicate with one another by passageways of plasmodesmata.

Besides having walls, plant cells have various other internal structures absent in animal cells. One of the most important structures are the vacuoles that were discussed in chapter 2. Vacuoles are the "water balloons" inside a plant cell that give the cell its turgidity. When water is sufficient, the vacuoles swell and push the cell membrane outward to the cell wall. When water is scarce, they deflate and the cell membrane recedes from the cell wall. When this happens, the plant wilts. Typically, plant cells have a single vacuole, and it stores water and various ions dissolved in it.

Plant cells have plastids; animal cells don't. Plastids store many significant compounds necessary for plants to function. The most significant of these plastids are the chloroplasts, which store chlorophyll. Not every plant cell is involved in photosynthesis, however, so not every plant cell contains chloroplasts. Chloroplasts are mostly found in places nearest to where sunlight strikes the plant, especially in the leaves and green stems. Cells in these areas often contain dozens of chloroplasts. Chlorophyll breaks down over time, so chloroplasts are involved in producing it too. As chlorophyll reflects green light instead of absorbing it, green plant tissues change color once its production is halted. Healthy green leaves and stems are the outward sign that the chloroplasts are functioning normally.

Plants store other compounds besides chlorophyll. Oils, starches, and proteins are stored in leucoplasts. These are not found uniformly throughout a plant either, but in cells specially devoted to this function. Seeds are one place that leucoplasts might be concentrated. The energy-rich compounds stored in leucoplasts fuel the growth of the embryo inside the seed. We consume seeds for the same nutritional value they provide us. Our dietary choices are made, in part, because of how plants distribute their leucoplasts.

Certain pigments are stored in chromoplasts. These pigments are essentially the same as the carotenoid pigments that support photosynthesis, but they have a different function. Chromoplasts are found mostly in fruits, flowers, and roots, and they are responsible for their distinctive colors. The color change we see as fruit ripens or flowers mature is the result of an increased accumulation of carotenoid pigments in chromoplasts.

Like chloroplasts, chromoplasts are responsible for producing carotenoid pigments as well as storing them. Chromoplasts synthesize a wide variety of pigments such as orange carotene, yellow xanthophylls, and various red pigments. The different flower colors we see in a row of zinnias (*Zinnia elegans*) or in the skins of a bunch of "rainbow" carrots (*Daucus*

The array of colors found in the petals of these zinnias (*Zinnia elegans*) is the result of carotenoid pigments that serve a variety of important plant functions.

carota) are the result of genetics and differences in the relative concentrations of these pigments.

It is easy to discern an evolutionary purpose for chromoplasts located in flowers and fruit. Bright flowers attract pollinators. There is no reason to attract them until the flowers are ready for pollination. Petals, while inside the bud, are rarely as brightly colored as when the bud unfurls. The color change can be amazingly rapid.

Many types of fruit are eaten and then dispersed by wildlife. This model is very effective once the seeds are mature enough to germinate, but it is lethal to the embryos in the seeds if consumed before. That is why most fruit are green and camouflaged by foliage when immature and become brightly colored when they are ripe. By advertising the presence of their ripe fruit, plants attract the animals that will consume and disperse their seeds. None of this relationship is haphazard. It is a dance whose steps have been carefully orchestrated over millennia. It is more difficult to provide a reason why various vegetables employ bright colors on the surface in their roots and tubers. What advantage a radish (*Raphanus raphanistrum*) might have by being red is a mystery yet to be solved.

Except for some specialized cells in the vascular system, plants are composed of three cell types. Nearly everything is made up of just one of these, parenchyma. Parenchyma cells are the simplest in structure, have a thin cell wall, and are designed to function as generalists. Parenchyma comprises the central portion of the stems, the inner layer of cells in a leaf, the material inside seeds, and the pulp of fruit. Parenchyma forms all the nonspecialized parts of a plant. The soft inner material in a stem, known as "pith," is comprised of parenchyma cells. With a thin cell wall, they are not rigid when the vacuoles gain or lose water, and their flexibility makes them ideal for storing things.

Plants store two other things besides water in their parenchyma cells: the engine that drives photosynthesis and the energy they need in the form of starches, oils, and proteins. There is no reason to put chloroplasts inside cells with thicker cell walls. Sunlight, water, and carbon dioxide need to reach them without any sort of hindrance. Parenchyma cells, with their thin cell walls, suit that purpose best.

Starches compose the primary storage unit for all the energy produced during photosynthesis. Solar energy is converted to simple sugars like glucose, and these can then be very easily converted back to energy needed by living and growing cells. Most of the energy produced by photosynthesis is used to fuel a plant's basic metabolism and growth, but healthy plants often produce more sugars than they need. Animals store this type of energy as fat; plants store most of it as complex carbohydrates like cellulose. Some of this is converted into proteins and oils. These surplus carbohydrates, oils, and proteins are mostly stored underground in roots and stems and then pulled back up into the above-ground portion of the plant when needed. This occurs when the above-ground portion of a plant is damaged. Without photosynthesis to generate the energy needed to produce new growth, the plant is reliant on stored energy to regenerate stems and leaves. Stored energy is also important to plants that stop growing for a portion of the year and resume later. Overwintering plants, for example, use their energy reserves to fuel their basic metabolism, stay alive, and then put new leaves out so they can resume photosynthesis. Carrots and radishes are classic cases of roots designed to store energy over winter, while potatoes (*Solanum tuberosum*) and yams (*Dioscorea alata*) are modified stems that function the same way. Parenchyma cells are ideal for this kind of storage.

The other main function of parenchyma cells is to produce new plant cells—not just more parenchyma, but every type of cell the plant needs. Unlike most animal cells, plant cells do not start life with a specialized purpose. As they are formed, they have the capacity to be any type of cell required by the plant. New cells are produced by special parenchyma cells, and their function is determined after they start to mature. If a plant is wounded, for example, new cells are rushed to the wound site to replace any of the cells that were damaged. This amazing flexibility allows plants to form new stems after the original ones are nipped off by a rabbit or pruning shears, to form new roots following transplanting to a new location in the landscape, and to form new bark and vascular tissue after the trunk is cut by a riding lawn mower or weed whacker.

The other two major cell types are specialized to provide more rigidity than is possible with parenchyma, and they differ in the way their cell walls are thickened. Collenchyma cells have a thickened primary wall. This makes it more rigid than the cell walls of parenchyma but still allows for some flexibility. Collenchyma cells often form long, thin fibers in herbaceous stems that still need to bend. The "strings" in a stalk of celery (*Apium graveolens*) or a green bean (*Phaseolus vulgaris*) pod are good examples of collenchyma cells. Leaf stalks (the petioles) are reinforced with it too. Collenchyma cells allow enough support for the leaf to be held at the proper angle to the sun and enough flexibility to bend in the breeze.

Areas of the plant that need the greatest support are reinforced with sclerenchyma cells. Sclerenchyma cells have a double cell wall, and the outer one is specially thickened. Sclerenchyma cells occur in places that

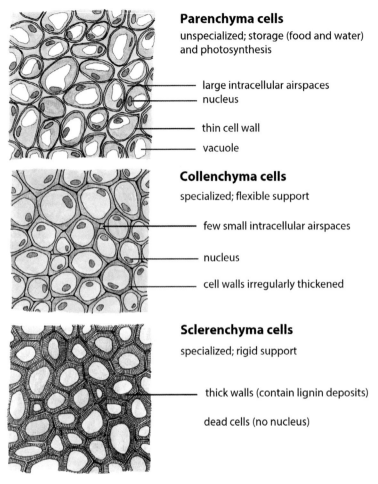

Parenchyma cells
unspecialized; storage (food and water)
and photosynthesis

— large intracellular airspaces
— nucleus

— thin cell wall
— vacuole

Collenchyma cells
specialized; flexible support

— few small intracellular airspaces

— nucleus

— cell walls irregularly thickened

Sclerenchyma cells
specialized; rigid support

— thick walls (contain lignin deposits)

dead cells (no nucleus)

Three cell types.

need extra strength and where having flexibility is less important. Whereas parenchyma and collenchyma remain alive at maturity, sclerenchyma cells are dead at maturity. Their sole purpose is reinforcement, and they do not have to be alive to provide that function. There are two different types of sclerenchyma cells. The type that forms long fibers reinforces the xylem and phloem that comprise the vascular system. Rope fibers are this type of sclerenchyma. The shorter, irregular-shaped sclerenchyma cells are used in places such as peach pits and the shells of pecans (*Carya illinoinensis*).

Almost all of the plant is composed of these fairly simple cell types, except within the vascular system. This network that functions as the plant's circulatory systems requires a few additional specialized cell types to work properly.

As discussed above, xylem cells carry water and dissolved nutrients from the roots to the top of the plant. They are stacked on top of one another to form an unbroken tube that runs from the root tips to the leaves at the upper reaches of the stem. These tubes are simple in structure, essentially hollow, arranged in bundles, and surrounded by a wall comprised of fibers that are normally sclerenchyma cells. The xylem of ferns and gymnosperms is composed of rather narrow tubes called tracheids. Flowering plants have tracheids too, but they have evolved larger-diameter tubes called "vessel elements." The larger tubes work much the same way but carry more water and do it more efficiently. Xylem cells die at maturity. In woody plants, this dead xylem becomes the wood. Our wooden furniture and the logs we burn in our fireplaces are composed almost exclusively of xylem cells—tracheids and vessel elements if it's oak or maple and only tracheids if it's pine.

The other type of specialized plant cell is responsible for carrying the sugars produced by photosynthesis as well as all the other plant-produced products (hormones, for example) and introduced pathogens. These products are dissolved in water that originally enters the plant via the xylem, but they are moved throughout the plant by a different system, the phloem. Phloem cells are also structured as tubes, and their design is similar in many respects to the tracheids and vessel elements discussed above. The tubes that comprise the phloem are called "sieve tube elements." They too are stacked end on end and form long "straws" throughout the plant. They also are arranged in discrete bundles inside the vascular system, and they are adjacent to the bundles of xylem, then surrounded by a sheath of fibers.

Sieve tube elements are normally a bit smaller in diameter than xylem tubes. This makes sense ecologically as it is more critical to the health of a plant to carry water rapidly than it is to transport sugars. Each sieve tube cell has a cap that looks somewhat like a manhole cover, the sieve tube plate, and it has a very specialized cell attached to its side, called a companion cell. Sieve tube elements remain alive after they reach maturity, but they are sort of like zombies; they do not have a functional nucleus to direct cellular activity or mitochondria to provide energy. These important functions are taken care of by the small companion cell attached to its side. No other cell in the plant or animal world is structured this way. Phloem is nearly identical in all vascular plants. Ferns and gymnosperms have less-developed sieve cells than flowering plants and lack companion cells and sieve tube plates. They function similarly, however.

Phloem has to carry its materials in all directions throughout the plant. Sugars produced in the chloroplasts enter the main phloem system at the capillary-sized ends of these tubes. It takes energy to pack these products across their cell membranes, and this energy originates from photosynthesis. Since every living cell in a plant is respiring, they require energy to fuel their basic metabolism as well. Simple sugars, such as glucose, provide this energy. Excess sugars get stored, mostly in the below-ground root system, and get transported downward, while the energy needed for life gets transported upward and downward to all the living cells.

Plants also take the simple sugars produced in photosynthesis and combine them with elements such as nitrogen to produce proteins and oils. Plant proteins and oils generally are stored for specialized needs such as fueling the development of their growing embryos inside their seeds. All of these plant-produced products are transported in the sieve tube elements of the phloem.

Plants have a highly developed hormonal system, equal to that of animals. Most of us understand the roles that hormones play in our life, for example, in the "fight or flight" reaction we experience that is fueled by adrenaline or the influence of testosterone in sex and aggression. Plants also have to react to the world around them and use a complex suite of hormones to do so. These hormones are carried in the two-way system of the phloem.

All the diseases and pathogens that affect plants also are carried in the phloem. Fungal and bacterial diseases in the soil enter the root system but are not carried in the xylem. Pathogens, like Dutch elm disease or bay laurel wilt, enter the vascular system through the saliva of sucking insects. These diseases are carried throughout the plant exclusively in the sieve tube elements.

PLANT TISSUES

As in animals, plants arrange these cells into tissues. There are three types of tissue in plants. Plants are mostly composed of ground tissue, and ground tissue is composed almost exclusively of parenchyma cells, though some collenchyma and sclerenchyma can be present for specialized purposes. Ground tissue carries out most of the basic functions required by plants, with support, photosynthesis, and reproduction being some of the most important.

Except in woody plants, dermal tissue is only one-cell thick and covers the outer parts of a plant—the roots, stems, leaves, fruit, etc. Dermal tissue is composed exclusively of parenchyma cells. As skin does in animals, dermal tissue provides protection to the inner parts of a plant from outside forces. It does not have chloroplasts or other organelles that might interfere with the movement of sunlight to the ground tissue just below it. For the most part, dermal tissue is extremely uncomplicated in design.

Dermal tissue, however, has several other important functions besides forming a simple barrier to the outside world. It produces "hairs" (properly called trichomes) of varying shapes and purposes on the surface of leaves, stems, and roots. Above-ground trichomes reduce water loss and herbivory. Dense mats of hairy trichomes often occur on the underside of leaves that reside in water-stressed environments. In deserts and other severely water-stressed places, they may occur on the upper leaf surface too. Such hairs on the upper surface impede sunlight from reaching the chloroplasts, but in the harshest environments there normally is plenty of sun but less water than desirable. Soft trichomes are also more often present on new growth than on older leaves. This is designed more to reduce the palatability of the foliage and stems to herbivores than to reduce water loss. Gathering a mouthful of leaves covered by "felt" is less palatable to most grazers than foliage not encumbered by it.

Trichomes also can be thorny. Spines and thorns on the surface of a leaf reduce herbivory even more. Such trichomes are more common in plants adapted to severely water-stressed environments than in less harsh localities. Having its stems and leaves chewed by herbivores not only impacts a plant's growth rate, but it also causes water loss. Just as most animals bleed from an open wound, plants lose water when damaged. Such wounds are healed as quickly as possible, but they bleed water until this happens. Desert plants especially cannot afford to lose water. Spines and thorns make it less likely they will be chewed upon. Having less-succulent foliage also minimizes this risk.

The specialized "hairs" produced on the stems and leaves of the various plants known as "stinging nettles" and poison ivy (*Toxicodendron radicans*)

The hair-like structures known as trichomes protect new growth from herbivores and reduce water loss from transpiration.

are trichomes as well. So are the trigger hairs inside the modified leaves produced by Venus flytraps (*Dionaea muscipula*).

At maturity, the epidermis of most plant roots produces root hairs to make the absorption of water more efficient. Root hairs are trichomes. Below-ground trichomes are less involved with reducing herbivory and primarily involved with acquiring water.

All trichomes are produced by dermal tissue. The relatively simple one-cell-thick layer continually works to produce and replace trichomes. In many plants, multiple kinds of trichomes are produced with varying roles. It is not uncommon for plants to have spiny and felt-like trichomes on different parts of their leaves and stems, for example.

Dermal tissue also produces a substance that covers the upper surface of most leaves and stems of terrestrial plants. Known as a cuticle, this substance makes various plant parts appear shiny or waxy and greatly reduces water loss. It also has some role in reducing herbivory and the potential damage of ultraviolet light. As with trichomes, plants that reside in severely water-stressed habitats generally have a thicker/waxier cuticle than plants in less stressful environments. Tropical plants, however, also have to deal with water loss during the peak sunlight hours. Many of them produce a shiny, thin cuticle on the upper surfaces of their leaves that retards water loss but does not greatly impede sunlight from reaching the chloroplasts.

The last major specialty of dermal tissue is the production of cells responsible for opening and closing the pores involved with gas and water exchange during photosynthesis, the stomata. Stomata are primarily found on the underside of leaves or the surface of photosynthetic stems. They are opened and closed by two cells that lie laterally to the hole, the guard cells. Guard cells are somewhat like water balloons. When the water pressure inside the plant is sufficient to maintain turgor, the guard cells are full of water, and they fall away from either side of the opening. When the water pressure inside the guard cells is reduced, the two guard cells collapse on one another, and the opening is closed. Water to fill the guard cells is controlled by a cell on the outside of each guard cell, the subsidiary cells. Subsidiary cells regulate the water balance of the guard cells. They send water into them to cause the stomata to open, and they hold water back to close them when water inside the plant gets scarce. The careful timing of this system is critical to the maintenance of water balance in the plant and its ability to photosynthesize.

I have already discussed the third tissue type, vascular tissue. The vascular tissue is equivalent to our circulatory system; it moves materials throughout the plant. There isn't, however, a central organ, like a heart, to create the movement. Ferns, gymnosperms, and flowering plants have a vascular system. Mosses and their relatives do not. Vascular tissue is composed mostly of the xylem and phloem, their specialized cells in a central core of tubes, often

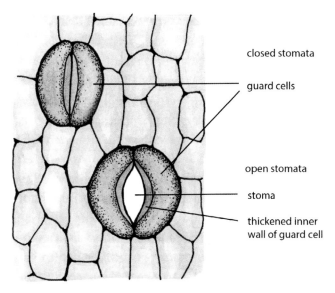

closed stomata

guard cells

open stomata

stoma

thickened inner
wall of guard cell

Detail of stomata.

surrounded by a wall composed of sclerenchyma or collenchyma fibers and with a few parenchyma cells for storage. This arrangement is often referred to as a "vascular bundle." Vascular bundles generally begin at the sites of photosynthesis and at the root tips and eventually form a complex network of ever-larger tubes throughout the main stems and branches. For it to work properly, the entire vascular system must be interconnected without gaps. The mycorrhizal network must be connected as well. Truly parasitic plants, like mistletoe (*Viscum album*), have to access this network to access the plant products they need to sustain themselves. This is done by injecting a hook-like structure, the haustorium, directly into the vascular bundle. Once this connection is made, the parasite can freely suck water and plant products from the xylem and phloem. Epiphytes, like Spanish moss (*Tillandsia usne-oides*), only attach themselves to the outside of the plant and do not create a connection to the host plant's vascular system. They live independently.

Plant organs are collections of all three tissue types. Though we might not think of them this way, they include flowers/fruits, roots, leaves, and stems. All of these major organs are composed of dermal, ground, and vascular tissues, and like animal organs, they drive significant functions of a plant's life.

The practice of "topping" a woody plant forces it to produce a new main leader to resume normal growth. Frequent topping saps energy from the plant that could have been used for growth and development.

Unlike animals, plant growth is essentially indeterminate; animals grow until they reach their mature size, and plants grow throughout their lifetime. Though the upward growth rates of stems and the downward growth rates of roots slow over the life of a plant, they do not cease completely. Each growing season, plants grow a bit larger at their tips. Apical growth typically has no end until the plant dies.

Lateral Growth

Lateral growth is different from apical growth. The increased girth of the main stem can occur in two very different ways: by producing true wood through secondary growth; or by simply accumulating old, dead leaf bases over time as the leaves form and die. Palms, bananas (*Musa* spp.), bamboo, and large cacti and succulents, like the Joshua tree (*Yucca brevifolia*), do the latter. They gain their ultimate girth as very young plants and then grow upward. Once they do this, their trunks do not get wider.

Palms are a unique case. When palms shed their leaves, they lose only the leaf blade. The petiole stays attached to the stem and layers of old leaf bases accumulate over time. This makes the palm stem wider even without having true secondary growth. Here in Florida, the native peoples and then the early European colonists used palm trunks to build lodgings that look much like the log cabins farther north. Such structures are not subject to the same type of decay as wood because the logs are really comprised of very tightly packed grass-like fibers. These fibers will rot from the inside out if the ends of the logs are not sealed but will last a very long time if only the outside of the trunk is exposed to the elements.

Secondary growth, the kind of lateral growth that occurs throughout the life of a plant, only occurs in woody species; gymnosperms such as the various conifers (pines, firs, spruces, cedars, etc.) and in flowering trees and shrubs. True secondary growth is also indeterminate; wood is laid down each growing season. Over the years, a woody plant gains girth in its trunk, its branches, and its roots. The rate at which this wood is added

Palms are not woody plants and do not produce lateral growth. These palm logs are fibrous and do not have growth rings.

varies greatly and is related to the plant's genetics and its growing conditions.

Secondary growth is almost exclusively composed of wood (xylem); a much lesser percentage is composed of bark and other layers. Large plants, such as trees and shrubs, require large amounts of xylem to carry the water they require. Each year, as they gain size, they lay down proportionally larger amounts of xylem.

Xylem dies after it matures but still acts to carry water and dissolved nutrients. A new layer of xylem is added each growing season, and the older xylem is squeezed inward into the core of the plant. Eventually, the center xylem becomes extremely dense and is often referred to as "heartwood." Organic compounds, such as lignin, are added to this core to give it extra strength and disease resistance.

In temperate regions of the world and in places where there is a distinct wet and dry period, plants have active and inactive periods of growth. Growth in such areas occurs in distinct pulses. Each dormant season, growth essentially comes to a stop. As nearly all of the lateral growth occurs in the wood (i.e., the xylem), this cessation of growth is most observable here. The lack of production of new xylem cells means that the distance between neighboring cells is extremely short, and this then appears as a ring. When growth is renewed, new xylem is added rapidly and it accumulates outside last dormant season's band. The amount of new xylem added each season is dependent on growth rate, and this is influenced by growing conditions. Scientists can infer a great deal about past climates by examining the rings of wood found in old or fossilized stumps.

In places like the wet tropics where growing conditions allow woody plants to grow year-round, rings do not form in such distinct patterns, but indistinct rings often still form if there are regular climate changes during the year. Periods of stress will slow growth, and the production of new xylem will also be reduced, making each layer of new cells narrower and appearing darker in color. Periods of good growing conditions will do the opposite. Though rings might not be readily apparent in these kinds of woody plants, a close examination of the wood can still yield important climate information.

As discussed above, tip growth is generated by the stem and root apical meristems; lateral growth has to come from different meristematic layers

True woody plants add girth annually. Those that live in areas with distinct seasons do so in spurts, which produce the annual rings that allow us to age them.

not present in other plants. Woody plants actually have two unique layers of meristem to produce the new cells they require.

Perhaps the most significant of these is the vascular meristem layer, known as the vascular cambium. As its name suggests, this single layer of cells occurs within the vascular tissue of each stem and root. It lies between the bands of xylem and phloem cells. As it produces new cells, the xylem is produced inside this layer, and new phloem is produced outside. As discussed above, the xylem is wood; the phloem remains a relatively narrow band of somewhat spongy material that is sometimes called "inner bark," though true bark is something completely different. The vascular cambium continually produces new xylem and phloem throughout the life of the plant and contributes most of the increasing girth that occurs in the trunk, branches, and roots.

The outer bark in a woody plant replaces the function of the single-cell-thick epidermis found in green plants. While the vascular cambium produces the wood that gives most of the strength required by the plant, the outer bark provides it protection from pests, disease, and weather phenomena. Different woody plants have very different types of bark. Each woody plant species has evolved its own set of strategies to cope with the types of environments it occurs in. Trees adapted to frequent fire often have very thick "corky" bark, whereas those that grow in less stressful environments often have thinner bark.

Outer bark is produced by the second lateral meristem layer, the cork cambium. This one-cell-thick layer of meristematic cells lies just under the bark itself. Over the life of a woody plant, it continues to produce new cells. The cells to the outside of this layer become new bark, whereas the ones on the inside become a thin layer of parenchyma cells known as the "cork parenchyma." The

The corky outer bark found in woody trees and shrubs protects the sensitive inner parts from disease and damage.

significant layer is the cork or bark. In some woody plants, this layer grows ever-thicker throughout the life of the plant, but often the outer/older layers are sloughed off periodically and replaced by the younger bark interior to it.

Look beneath most old trees and you can find discarded bark that has been replaced. The ability of woody plants to generate new bark protects them from the damage caused naturally by pests and extreme weather phenomena. It also allows us to harvest it for our own uses without seriously damaging the plant, as long as we don't harvest so deeply that we damage the cork cambium layer itself. Cinnamon (*Cinnamomum zeylanicum*) is harvested this way in places like Sri Lanka; the outer bark is peeled away in thin sheets to make cinnamon sticks, and the tree is then allowed time to replace it. Actual cork is the outer bark of the cork oak (*Quercus suber*), a tree that is native to Mediterranean regions of Europe and North Africa. Having its outer bark removed this way lowers the plant's ability to withstand various external dangers and harvesting it has to be done carefully if the plant is to be protected and used in the future. Such things cannot be done, however, with nonwoody plants because they don't have a cork cambium layer to continually produce new bark.

CONDITIONS AFFECTING GROWTH

Overall plant growth is predicated on a complex equation that involves the plant's genetics and its growing conditions, and each of these involves multiple factors such as light, moisture, temperature, and nutrients in the outer environment as well as hormones in its inner environment.

When everything around a plant is optimal for its growth, its ultimate growth will be determined by its genetics. Just as we are a product of our mother's and father's genes, plants are too. We can give the same two plants the same growing environment in our landscape, but they are unlikely to respond identically if they were seed produced. Many plants in our modern era, however, are produced by stem cuttings or tissue culture. Such plants provide more uniformity in the landscape but may not be as able to defend themselves against pests and disease due to their lack of genetic diversity. In the wild, genetic diversity normally protects populations from universal damage. A rabbit, for example, may feed heavily on some of the population but find others less palatable. I often see this in my own landscape. Clones of a single plant have no genetic diversity to deal with these kinds of out-side stresses should they be exposed to them. Sometimes the problem never appears, and the clone is just fine. Sometimes a disease-resistant clone is actually preferable, at least in the short term. Maintaining genetic diversity also means that no two plants of the same species will look identical. Some gardeners prefer individuality; others prefer uniformity. If you are one of the

Seed-grown plants are never identical genetically and will show differences in growth and development.

latter, you are likely to be more satisfied purchasing plants that were grown from cuttings or tissue culture, not plants grown from seed.

The expression of a plant's genes is influenced internally by a plant's hormonal system. Plants that are "relaxed" will grow differently than those that are stressed. It's no different for plants than it is for animals. The environment we put a plant into will affect its hormonal system and its ultimate ability to grow to its full potential. This goes beyond simply meeting its light, water, and nutrient needs. Plant hormones will be discussed in greater detail in chapter 10.

Plants take a conservative approach to growth. Not all of the energy gained from photosynthesis is put into growing. Some of the energy is stored to provide for the plant in case times become leaner in the future. Just as we balance our personal finances into spending and savings accounts, plants must do the same if they are to survive such things as extended cloudy or chilly days that would limit photosynthesis or unfortunate accidents like having their top knocked down during a storm.

The energy provided through photosynthesis is used for everything from fueling the plant's core metabolic functions to moving sugars into storage, reproducing, communicating, recovering from disasters, and fighting disease and pests. This is really no different than it is for animals. We all try to save something for a "rainy day."

Unlike animals, however, plants adjust their basic metabolic rate on a daily basis. They are frugal; plants spend less energy on their metabolism than they earn each day, and they put something aside even on days when they earn very little. Plant biologists have found that plants have an internal mechanism that calculates daily the amount of energy they've gained and divides it by the length of the night. This solves the problem of how to portion out energy reserves during the night so that the plant can maintain itself, yet not risk burning off all its energy reserves before starting a new day.

The calculations plants make are extremely precise. Regardless of the length of daylight, plants recognize the amount of stored starch and adjust their metabolism to keep about 5 percent intact within their stores. Plants use this stored energy only when all other options are exhausted. Most can survive a relatively long time on their stored energy if kept in situations where their ability to photosynthesize does not meet their lowest metabolic needs. Eventually, they will use these stores up and starve to death.

LIFESTYLES

There are some things plants cannot effectively adjust—their basic lifestyle and their location once they have taken root. Therefore, it is extremely important that they take root in a location that fits their ecology. Evolution has shaped these choices for hundreds of thousands of years, and it has worked very well.

Plants have to make choices regarding their basic longevity. Just as some animals are exceedingly long-lived and others are not, plants make similar choices to adjust to the environmental conditions they live in. As for all living things, the overriding biological need is to live long enough to reach maturity and reproduce effectively. Everything else is frosting on the cake, figuratively speaking.

Many plants complete their entire life cycle in one year or less. Such plants are annuals. Annuals emerge from their seed or spore, reach maturity, produce new seeds or spores in one year or less, and then die. Such a life cycle is sensible for plants adapted to uncertain surroundings—not regular and predictable disturbances like annual cold but unpredictable problems. If a plant cannot predict when the next calamity might occur, but lives in an environment where calamities are highly likely, it makes sense to complete its life cycle in the shortest time possible. Many annuals are weeds, for example. Being "weedy" has nothing to do with a plant's desirability and everything to do with its life cycle. Weeds grow and mature quickly, and they take advantage of disturbed locations before other plants can colonize them. Disturb an existing vegetated area, and what colonizes it first are mostly weedy annuals.

Annuals also have to have great faith in the future, given that they put all their seed out into the world at one time and die. If they were like many other kinds of plants, they might go extinct in the next year or two if conditions proved unfavorable for their seeds to germinate. Most annuals do two things to counter this: they produce huge numbers of seeds, and their seeds remain capable of germinating (i.e., remain viable) for years after they are shed.

Biennials are really annuals that take two years to complete their life cycle. Many of these come from environments that are so harsh or have such shortened growing seasons that it is extremely difficult to mature and set seed in one season. Biennials spend their first year establishing themselves

Weeds are simply plants that exhibit certain growth characteristics, in-cluding an ability to rapidly colonize bare and disturbed soils. Some, like these Queen Anne's lace (*Daucus carota*), can be quite attractive and desirable in a landscape.

and getting poised to reproduce the second season. Many of our most favor-ite root vegetables—carrots, radishes, beets (*Beta vulgaris*), turnips (*Brassica rapa* subsp. *rapa*), and parsnips (*Pastinaca sativa*) are biennials that have evolved under these kinds of growing conditions. The roots that we con-sume are storage roots, designed to store large amounts of carbohydrates over winter so that the plant can have a burst of energy in the spring and complete its life cycle before winter sets in again. If you've ever gardened with these vegetables, you've likely learned two things: the plants do not set seed the first year; and the roots do not stay edible in the second season. I learned this as a very young gardener one year when I did not pull all the radishes from my garden in the hope that they would become even larger the next spring. Much to my disappointment, the roots on my second-year radishes were thin and woody, not round and succulent.

Biennials flower only once, and the plant dies once the seeds are mature. Because of this, they also adopt much the same strategies as annuals. Both biennials and annuals need to be able to effectively reseed if they are used in a landscape, or they must be replanted constantly. In my wildflower garden at home, this means that I reduce the amount of mulch in my planting beds so that their seed makes contact with the soil. It also means that I can't remove the spent flower heads until I'm sure the seed has been dispersed.

Perennials do not live forever. Even the oldest plants on earth eventually succumb to old age. There are short-lived perennials and long-lived ones. Being "perennial" only means that reproduction occurs over a period longer than one year. Here in Florida, for example, our native sky-blue lupine (*Lupinus diffusus*) spends its first year as a small plant, produces a small number of flowers the second year, and then becomes a robust adult that bursts into bloom in year three. Once the seed pods have ripened sufficiently, this seemingly healthy three-year-old plant dies. At the other extreme are long-lived trees that may live through hundreds of reproductive events in their lifetimes.

Perennials generally mature more slowly than biennials and annuals. They tend to live in habitats where they can take their time reaching sexual maturity. Often they produce fewer seeds or spores per event than the other two types, and their seed doesn't have to remain viable for long periods. If things don't go according to plan this year, they have more opportunities in the future. Of course, there are many exceptions, and short-lived perennials, like the sky-blue lupine, often behave more like annuals and biennials than more long-lived perennials. On average, the seed of perennials should be planted within a year or less after it has matured.

Plants also have to choose whether to hold their leaves throughout the year or to shed and replace them on some kind of predictable cycle. Evergreen plants, like pines and southern magnolias (*Magnolia grandiflora*), do not shed their leaves at one time seasonally, but it is a bit of a misnomer to say they are "evergreen." All leaves wear out, get shed, and then get replaced. If you've ever

Biennials, like this honeycombhead (*Balduina angustifolia*), spend their first year growing, flower in their second year, and then die after their seeds are shed.

lived with a pine in your landscape, you know that it sheds its needles constantly, not all at once. Being evergreen makes sense in several vastly different kinds of habitats. In tropical areas where growing conditions are favorable year-round, being evergreen allows a plant to take advantage of this and never rest. In nutrient-poor locations, it makes sense because leaves don't have to be replaced in large numbers at any one time. In exceptionally cold habitats, being evergreen allows you to get a running start immediately after the spring thaw. Deciduous plants have to spend the first few weeks of spring replacing leaves before they can once again start photosynthesizing. Evergreen plants do not. It is not a coincidence that the percentage of evergreen trees and shrubs in a plant community typically increases in higher elevations and latitudes.

Plants that lose all their leaves at one time generally live in environments where growth during a predictable period each year is impossible or not sensible. As leaves are the organ by which plants photosynthesize and transpire, maintaining them over winter or during periods of drought is dangerous. That is why most woody plants adapted to cold winters or predictable dry seasons are deciduous.

No plant photosynthesizes when the ground is frozen as the roots cannot effectively supply the water that is needed. Evergreens keep their leaves,

Deciduous plants shed their leaves annually in response to a change in growing conditions such as freezing temperatures or seasonal drought.

but they have to protect them from the cold. They do this by producing compounds, such as terpinols, that essentially operate like an antifreeze. Deciduous trees and shrubs do not produce these types of compounds. The water in their leaves freezes when exposed to extreme cold and the tubes of xylem and phloem burst, just as a can of a carbonated beverage does when put in the freezer. Frozen leaves cannot function, and the damage cannot be repaired. If this occurs during a sudden and unexpected cold snap, new leaves have to be produced to replace those that have been irreparably damaged. Otherwise, plants prepare for the upcoming winter and shed their leaves before such damage can occur.

Plants that reside in places with distinct dry seasons also prepare for this by shedding their leaves ahead of time. Not all tropical areas, for example, are routinely moist. The so-called dry tropics have distinct wet and dry seasons. Maintaining leaves during extended drought puts excessive stress on a plant because plants lose up to 95 percent of the water they take in during photosynthesis. Maintaining leaves is not cost-effective at these times, and it risks the life of the plant. It makes far more sense to simply replace the leaves once the rains return. Many desert plants are leafless most of the year and produce new leaves for a brief period following a rain. Others simply never bother to produce leaves at all and use their stems for photosynthesis. Water is a precious commodity. Plants often conserve it by dropping their leaves during periods of drought and replacing them once the drought ends. Even plants not adapted to predictable drought will shed their leaves during extended dry periods and survive for some time afterward before finally expiring.

Deciduous plants risk far less by simply resting during predictable periods of stress than they do by trying to extend their growing season. Such plants may hold their leaves a bit longer during years when good growing conditions are extended, but they will eventually lose them even without a freeze or a long drought because that is the way they are programmed genetically. Being deciduous requires plants to be in tune to both their environment and their internal biological system.

5

Roots

Roots may well be the most important plant organ and the least understood. As the organ solely responsible for water and nutrient uptake, it is absolutely critical that a plant's root system remain in good condition and be allowed to function at peak performance, yet the fact that it exists largely out of sight precludes us from adequately assessing its overall health and often puts it out of mind.

The ultimate factor in predicting the long-term success of a newly acquired plant is the condition of its roots, but most of us purchase plants based on what we see above the soil line. We look at the strength of the main stem, the condition of the foliage, and the plant's overall aspect, but we rarely look at the condition of the roots by carefully removing the root ball from the pot and examining its growth and condition. Once planted, however, it will be the roots far more than the top that allows the plant to make an unhindered adjustment to its new home. A healthy root system will generate a healthy top, while a healthy crown will take far longer to regenerate a healthy root system that is currently unhealthy or inadequate. Ignoring the roots may well have significant consequences to your gardening experience.

Healthy roots are noticeably different from those that are stressed or declining. Roots that are actively growing produce an obvious network of bright-white feeder roots, and these will be encircling the edge of the container as well as in the interior. These thin, spider-like roots are the ones doing the bulk of the water and nutrient absorption. When roots start to decline, these are the first to go. Caught in time and given the proper environment, these roots can regenerate, but the plant won't recover if something isn't quickly done to stop the decline. This often catches us off guard because it can take weeks for the appearance of the top to show the same ill health.

The other aspect of roots to consider before making a purchase is how they appear within the container: are they fully scattered throughout the container, or are they wound around the sides in a tangled mess? Like all plant organs, roots have memory. As they grow and develop, they remain in the position in which they were formed. New roots will seek out new places to grow, but the portion behind will remain just as it is. This is not as important for herbaceous plants as it is for woody ones because the roots of herbaceous plants do not increase in girth. The roots of woody plants, however, do increase in girth as they age just like their stems do. If the roots of a woody plant have wound around one another in the container, they will

Roots produced in a containerized plant are forced to grow unnaturally in a circle. This is problematic in woody plants as their roots retain this position forever.

remain that way after planting *and* they will increase in diameter over time. Eventually, these ever-widening roots will essentially strangle one another, and this will significantly impede growth. Different specimens of the same tree or shrub will often show differences in growth rate over time even when they were of identical size at planting. Much of this can be attributed to differences in the original root balls.

Mosses and their relatives don't have roots; they are nonvascular. Ferns and their relatives begin growing their roots as soon as their spores germinate. In seed plants, the roots begin to form as soon as the ovule is fertilized and the embryo comes into existence. The developing plant embryo lies inside the seed, but even at its earliest development it is possible to detect the root apical meristem—the location where root cells will be produced as the embryo prepares to sprout.

ROOT STRUCTURE

Plant roots are comprised of all three tissue types. Their dermal tissue is one cell thick in nonwoody plants but is composed of bark in woody species. Mature root dermal cells produce trichomes; in roots these are the root hairs. The vascular tissue is composed of xylem and phloem, and the ground tissue is mostly parenchyma cells. As parenchyma cells are storage cells, this allows them to store the excess energy produced by photosynthesis.

When a seed sprouts, a single root begins the process of initiating the development of the overall organ. In most plants, this root burrows downward, extending itself to anchor the seedling and then branching in various directions before growing the root hairs that are vital in absorbing water and nutrients. A mature root system is extremely large in comparison to the size of a mature plant; a four-month-old rye plant (*Secale cereale*), for example, was estimated to have 14 billion root hairs, with an absorbing surface of more than 4,300 ft^2 (401 m^2), and an end-to-end length of more than 6,200 miles (10,000 km).

Regardless of the type of plant and its final root structure, this is essentially true. Though some plants, like certain mangroves, are specially adapted to produce roots that grow upward, this occurs later, after the initial root system is established. Most plants have specialized cells at the root tip called amyloplasts that permit the root to detect gravity, a tendency known as gravitropism. Just as we can detect up from down, even with our eyes closed, plant roots do the same. Without these sensors, new root growth is halted until they can be replaced. It is vital to a plant's success to know up from down before investing in additional root growth.

Roots have well-defined zones. The portion that contains the actively growing cells is the apical meristem. Like all meristematic tissues, the root apical meristem generates cells that are undifferentiated. As they mature, they will become a specific cell type and assume a specific role, but as they are generated by the root apical meristem, they are simply new cells with the capacity to become anything the plant needs.

The area immediately adjacent to the apical meristem is rather indistinct but continues to produce new cells, though a bit more slowly. Botanists divide these indistinct regions into two separate layers. The quiescent center is a population of cells that reproduces much more slowly than other meristematic cells and is resistant to radiation and chemical damage. These cells are possibly a reserve that can be called into action if the apical meristem becomes damaged. The zone of cell division includes three areas where the new cells initially differentiate into their new roles. The protoderm is the outermost primary meristem and produces cells that will become dermal tissue; the ground meristem is central and produces cells that will become ground tissue; and the procambium is the innermost primary meristem and produces cells that will become vascular tissue.

Once these cells are formed, they begin to elongate and grow. This "zone of elongation" is the region that actually pushes the new root downward or forward. Their cell walls are elastic, and they lengthen up to ten times their original size. The cells in this region, however, are still immature and incapable of absorbing water and nutrients.

Once the cells have stopped elongating, they become fully mature and begin functioning as roots. This "zone of maturation" is where each of the various cell types assumes its role as vascular, ground, or dermal tissue. It is the zone where the dermal tissues begin forming root hairs. Prior to this, there was no fully formed xylem to carry the water and dissolved nutrients. It is also at this stage that relationships are formed with various soil organisms, such as mycorrhizal fungi, to increase their efficiency.

The apical meristem also produces root cells that precede it. One sig-

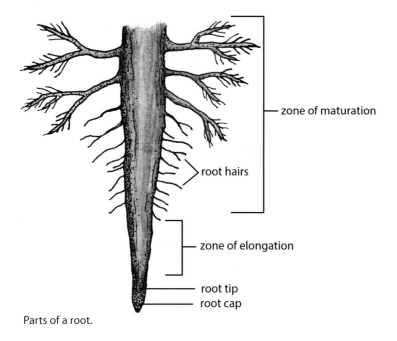

Parts of a root.

nificant region is the root cap. As its name implies, the root cap forms a protective covering over the sensitive apical meristem. Root growth through sharply edged soil particles is a contact sport, and as any sports fan knows, contact sports are much safer wearing a helmet. Root growth and development are dependent on the apical meristem remaining healthy and intact. The root cap absorbs the various shocks and abrasions as the root extends its way through the soil. While it does this, it becomes worn down, but the apical meristem replaces those cells to make sure it remains fully protected.

The root cap also is the part of the root that contains the amyloplasts, the cells that provide the root a sense of direction. The apical meristem is busy making new root cells. The root cap is in charge of housing the parenchyma cells that provide gravitropism. If damaged, the root apical meristem will replace all of this root cap structure, and all of this becomes the first order of business if damage occurs. If you transplant a plant from one area of your landscape to another, many of the root tips will be damaged in the process. Damaged roots first regenerate their apical meristem. Then, before any new growth can occur, they produce new root caps. There is no sense sticking your neck out into new territory without first having a protective covering on your head and a road map to help guide you.

Burrowing through soil particles produces a lot of friction. The root cap solves some of that by producing lubricants that reduce it, just like engine oil in a motor. The root cap cells secrete mucilage. This is a polysaccharide substance that absorbs moisture from the soil and expands into a slimy lubricant. There is some disagreement regarding the overall importance of these mucilages, but plant researchers have shown that they do, in fact, increase a root's ability to penetrate the soil. Mucilage seems to be less effective in dry soils, but it clearly helps in compacted soils.

The second significant area formed by the root apical meristem and preceding the root tip is the region of border cells. The role that border cells play has only recently been examined and is the subject of much current research. Border cells actually slough off the root tip during root elongation. They are totally detached from the root cap but remain in close communication with it as the roots burrow through the soil. In a way, they act as scouts, working ahead of the main root system, looking for dangers and opportunities and helping to direct the root's forward progress.

Root border cell production is enhanced by increased soil compaction. The whole of the root cap surface may be covered by these border cells in densely compacted soil, and they reduce the friction between the root cap surface and surrounding soil particles. In this way, border cells help to augment the role of the root cap.

Once the border cells are sloughed off the root tip, they remain close by and are kept lubricated by the mucilage generated by the root cap. It is this mucilage bath that allows the border cells to remain in contact with the root system and provide the other functions so necessary for its overall health.

Border cells play a significant role in enhancing the root's ability to absorb water and nutrients. Besides mucilage, border cells secrete a chemically complex bath of plant metabolites, proteins, and extracellular genetic compounds that actually attract the types of mycorrhizal fungi necessary for the roots to function effectively. Plant roots can't rely solely on chance for this necessary relationship to develop. The secretions of the border cells signal the impending arrival of new roots to these fungi, and this draws the fungi to the root's location. Without border cells, this vital lifetime relationship might never be forged.

In addition to attracting beneficial fungi to the developing roots, different secretions of the border cells provide antimicrobial properties that protect them from a wide variety of soil pathogens including harmful fungi, bacteria, and nematodes. Seedlings exposed to pathogenic fungi often suffer extensive damage to the zone of elongation, but little damage to the mature portions of the root and the apical meristem. A protective mantle, generated by the border cells, effectively shields the most sensitive portions of the roots from damage, and this allows them to eventually recover.

Border cells also reduce the damage that high concentrations of certain soil chemicals produce. In the presence of high aluminum levels, a situation common in low pH soils, root border cells produce increased amounts of mucilage. Laboratory experiments have shown that this thicker mucilage layer increased the survival of seedling plants and enhanced their productivity, protecting the root from pests and disease.

HOW ROOTS WORK

The basic design of a plant's root is rather simple. The way it functions is a bit more complex. Water with its dissolved nutrients is attracted to the root's dermal tissue because plant cells have a wall composed of cellulose. Cellulose acts like a sponge and is the reason your paper towel or tissue works the way it does. Once inside, the water has to travel all the way to the xylem in order to be transported up into the upper branches and leaves. As discussed above, moving water through the plant does not require energy; it is done by transpiration. Water lost through the open stomata creates a pressure gradient that sucks water up like a straw.

Water enters near the tip of a growing root through the root hairs, but this is greatly facilitated by their relationship with mycorrhizal fungi because the fungi permit the root hairs to be in closer contact with the soil particles than they would be otherwise. Once past the epidermis and inside the root itself, water reaches the xylem by first going through the outer layer of parenchyma cells that comprise the root cortex. The pull generated by transpiration draws that water toward the xylem bundles in two ways. A small portion is actually pulled through the cortex cells themselves (the symplast pathway), but most is pulled around the outside of these cells through the air spaces between them (the apoplast system).

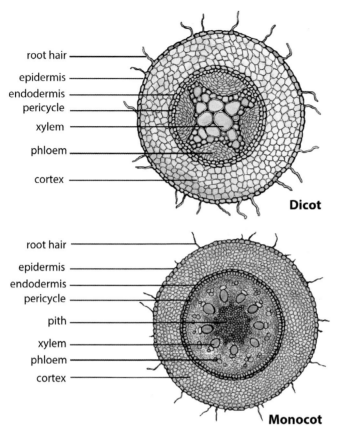

root hair
epidermis
endodermis
pericycle
xylem
phloem
cortex

Dicot

root hair
epidermis
endodermis
pericycle
pith
xylem
phloem
cortex

Monocot

Root cross section.

Water then reaches the outer edge of the vascular bundle that contains the xylem and finds its first impediment, the endodermis with its Casparian strip. The endodermis is a single-cell-thick line of cells that encircles the vascular bundle. Woven through it is a waxy strip known as the Casparian strip. Casparian strips are made of lignin, the same polymer that is used for building the xylem tracheids and vessel elements and the material that is added to old xylem to make inner wood so strong and pest resistant.

Casparian strips provide an extracellular diffusion barrier within the plant roots, forcing nutrients to pass into the cells, where they are then subjected to the action of the transport proteins on the surface of the cell membranes. This is an "apoplastic" barrier since it forces the water and dissolved minerals to go directly through the plasma membranes of the endodermal cells. By this method, plants have some control over the amount and type of nutrients they let into the xylem.

The endodermis is essentially a gatekeeper, making decisions as to what will pass into the xylem and what will be left outside in the root cortex. Water always passes by unobstructed, but the dissolved minerals are carefully screened. Plasma membranes are exceptionally complex structures and well beyond the scope of this book. Simply described, they allow water to pass through, but they prevent the minerals dissolved in the water to pass without carrying them across using certain protein molecules. These "carrier proteins" attach to the surface of the plasma membrane, and it takes various enzymes to take them from the membrane surface to the inside of the cell.

By this system, plants have the mechanism they need to regulate their nutrient intake. The plasma membrane surrounding each cell of the endodermis, coupled with its Casparian strip, maintains the proper balance of minerals within the plant while allowing an unimpeded flow of water.

Types of plants differ in how their root systems are structured, but the mechanics of how they work are the same. Ferns and their relatives have very primitive roots. Tough fibrous roots grow off their rhizomes (their stems), and they branch sparingly or not at all. Because ferns roots develop from the main stem and not from a primary root produced by the embryo, they are called "adventitious" roots. This makes dividing a fern much different than dividing most other plants as any reasonably sized piece of its rhizome will include some roots as well as a frond. Since fern roots are unbranched and wiry, it is relatively easy to disturb them without the same kind of damage that would occur in other vascular plants.

Gymnosperms produce a taproot as do all of the flowering plants not classified as monocots—the dicots, magnoliids, and the basal angiosperms. A taproot develops as a single main root at germination and continues to develop over the life of the plant. Carrots are good visual examples of taproots. As a taproot buries itself into the soil, it anchors the juvenile plant and keeps it upright so that the rest of its growth continues unimpeded. Taproots do not grow forever, and, in most plants, the bulk of the future growth occurs as branches that develop off the taproot. Though the taproot provides the best means of anchoring a young plant, it is very ineffective in providing the stability needed by a large tree or shrub. Engineers know this when they design large bridges and skyscrapers. The weight load needs to be distributed laterally, not horizontally, for such expansive structures to withstand the forces to which they will be subjected. It's the same with plants.

Once the juvenile plant has anchored itself with its taproot, it begins the process of developing a broad lateral root system to support its future growth. These roots—not the taproot—form the base that will protect the mature plant from falling over in high winds and saturated soils. It is easy to see this structure when a tree topples. It also is apparent if you've ever tried to add a new plant to the soil beneath the canopy of a mature shade tree.

This lateral root system develops early and is the source of so many problems in container-grown trees and shrubs. When woody plants are confined too long in a container, their taproot will quickly reach the bottom and then have nowhere to go except sideways or upward, coiling around itself as it does so. That is a problem, but a bigger problem is created as the lateral roots develop in this container off the main taproot. It's the lateral root system that provides much of the stability to a tree or shrub and most of the nutrient uptake. Creating problems with this system will have long-term negative impacts on these plants. If they are the roots of woody plants, this spiraling and coiling effect will exacerbate the impact to the plant as each of these roots gains girth over the years. Herbaceous plants, like perennial wildflowers, don't have woody roots and will recover from this effect if planted properly and cared for long enough. Woody plants will never recover unless some root pruning is done first.

If the roots in your container-grown plant have not become too coiled and tangled, try to untangle them as much as possible at planting time. This, of course, is a bit subjective. If the root systems cannot be easily uncoiled and the main roots have already become large and woody, the roots should be cleanly cut to remove the most obvious problems. Remove the bottom ½–1 inches of the root ball with a sharp clean shears and ½ inch along the sides. Remember that roots are the primary organ required by the plant for water and nutrient uptake and that removing some of this system will make it more difficult for the plant to provide for itself. Root pruning hurts its short-term

Typical roots of a dicot plant develop early once the initial taproot is formed. Branching roots form the structures most responsible for stabilizing the plant and producing its feeding network.

growth but releases it from long-term damage. Take extra care to water root-pruned plants long enough to give them time to redevelop a full root system and make sure they have ample fertility during this time. In nutrient-poor soil, it is helpful to add some extra fertility with either an organic fertilizer or an inorganic one specially formulated for this purpose. Such fertilizers will be very low in nitrogen and much higher in potassium and phosphorus.

The same issues need to be addressed when working around established trees and shrubs with this type of root system. Trenching and digging in the soil near trees will cut roots and damage the tree, causing a general decline in its vigor and/or making it more susceptible to falling over in a high wind or waterlogged soil. Cutting roots greater than about one inch in diameter during trenching and digging can mean problems for the tree as these are the major structural roots holding the tree upright.

The impact from pruning roots depends on several factors. It is more injurious to old, mature trees than to younger, more vigorous ones. Damage typically increases with more cuts, bigger cuts, and cuts made closer to the trunk. Root pruning, trenching, and other construction activities close to the trunk result in more injury on shallow, compacted soils or on soils that drain poorly than on well drained soils. Trees that are already leaning are poor candidates for root pruning. For mature trees, most experts recommend not cutting roots closer than 6–8 inches from the trunk for each inch of trunk diameter.

Monocots (grasses, palms, orchids, lilies, etc.) have a very different root structure than gymnosperms and the other types of flowering plants. Monocots have a single root that emerges from the sprouting seed, but it soon disintegrates once the seedling is established and is usurped by a region of root-producing tissue at the base of the stem. By definition, these are adventitious roots, but in monocots these are typically referred to as a fibrous root system.

Calling them "fibrous" is an apt term because they rarely, if ever, branch. Hundreds to thousands of these single roots emerge from the base of monocot plants, and they reach a length dictated by the plant's ecology. They grow outward and downward as a thickened brush.

Few monocots attain the size of woody trees and shrubs due to this change in engineering design. Those that reach significant heights, palms for example, accomplish it by developing a very dense system of thickened root fibers and a trunk that is extremely flexible so that some of the energy that blasts it during an extreme storm is absorbed by the above-ground portion instead of just the root system. They also develop far fewer spreading branches, like those of an oak or maple, which would make them top heavy. The nearly uniform basic design of palm trees has been dictated by its root structure.

Monocots, like this grass, produce a fibrous network of roots from a main node beneath the leaves.

Because of their fibrous root system, monocots suffer few consequences from being confined in a container. Monocots never produce wood, so their roots do not gain girth over time, and all the new root growth that occurs after planting will arise as new roots from the base of the plant, not from a lateral root off an existing already twisted tap root. Container-grown monocot roots may not be able to reach their full length while containerized, but this is fairly easily rectified once the plant is removed from the container and established in the landscape. Palms, grasses, bamboos, and the like can be grown in relatively small containers and not have it affect their future growth and survival.

This also allows plants like these to be root pruned and/or transplanted with relative ease. Digging around a palm to install a new plant or during construction will not cause anywhere near the same type of consequences it would to a woody tree or shrub. Even removing the plant for transplant purposes can be done rather easily as long as care is taken not to damage the base of the stem where the production of new roots occurs. Here in Florida, it is common to see mature palms being transported on the highway on the bed of large trucks with only a tiny root ball at the base of the tree. This could never be done with a mature pine or oak. As long as the palm is staked long enough for a new fibrous root system to develop from its base and some of its fronds are trimmed to reduce water loss from transpiration, it will recover and eventually thrive.

Though taproots and fibrous roots differ somewhat in structure, the vast majority of roots are found relatively close to the soil surface. In deserts and other arid environments, some plants develop deep root systems to find water and dissolved nutrients, but these are exceptions, and even desert plants form extensive root systems near the soil surface to catch dew and other moisture before it can evaporate. These so-called feeder root systems are designed to capture water and minerals where they are most easily acquired; near the surface where organic matter is decomposing and where water is first falling. It follows the laws of competition that plants will try to capture what they can before others have a chance to capture it too.

When water falls on the soil surface, it picks up essential minerals and begins to move deeper into the soil column. As it does so, the roots of every plant in the plant community compete for it. Eventually, the nutrient-laden water moves so deeply into the soil column that it can't be reached at all. It's this competition within the root zone that dictates that plants be most aggressive near the soil surface before others have a chance to take it for themselves. Competition for water and nutrients occurs throughout the root zone, but it is most aggressive in the upper few feet. This also is the region where most of the beneficial soil organisms reside. In healthy, not-too-droughty soil, the upper few feet of the soil column has the most active root network, and this should be protected as much as possible to protect the health of your plants. The large surface roots that are often noticeable around large trees are the structural roots that help keep the tree from tipping over, but the job of absorbing water and nutrients is occurring mostly at the root tips, and the bulk of these are near the surface, not six or more feet below.

Not all plants position their roots below ground. Some, like this orchid, attach themselves above the ground on other plants (epiphytes) or on rocks (lithophytes) and are especially adapted to get the water and nutrients they need from rainwater and dew.

Various types of plants can develop specialized roots that relate to their specialized ecological needs. Plants that grow on the sides of taller plants or on rocks (epiphytes and lithophytes, respectively) produce roots specialized in holding the plant firm to its host while obtaining sufficient moisture in these locations to meet their needs. Epiphytic and lithophytic orchids do this by having a spongy covering on their roots called vellum that is superabsorbent and capable of absorbing water vapor from mist in the air. Many vines and lianas climb into the canopy by means of sucker-like pads or hooks at the end of their climbing roots. These types of roots are structured a bit differently than the ones that develop in the soil below, but they are capable of absorbing water and nutrients in the rainwater that washes over them. Since these types of roots arise on the stem, they are adventitious.

Many trees and shrubs adapted to saturated soils, like cypress, develop above-ground structures (e.g., knees and the like) to assist them in respiration or buttressing roots to improve their stability. Many wetland plants, woody and herbaceous, develop adventitious roots just above the waterline for the same purpose if they are flooded for extended periods. Few plants are actually parasitic, like the mistletoe, and develop specialized roots that burrow into the vascular systems of their hosts. Specialized roots evolve to perform specialized functions, but their primary functions and their internal structure remain basically the same.

6

Stems

Stems are the plant organ that connects the roots to the leaves. Stems come in all kinds of shapes and sizes, but their basic function remains the same. To perform it, they have to provide structure to the plant and a transportation pathway that allows for the free flow of water to the leaves and sugars to the roots and the rest of the plant. They also often play a significant role in photosynthesis.

Stems can be herbaceous or woody, but all stems contain all three tissue types; dermal, ground, and vascular. Dermal tissue is one cell-layer thick in herbaceous plants. This arrangement may skimp a bit on protecting the stem from environmental damage and herbivory, but it compensates by allowing for photosynthesis. Herbaceous plants supplement the photosynthesis that occurs in their leaves with additional chloroplasts just beneath the epidermis in the stem cortex. For sunlight to reach these chloroplasts effectively, the epidermis must be as thin as possible. Green stems represent a large fraction of the total photosynthetic area of most species, but they are less capable than leaves of adjusting to light. They have biomechanical and hydraulic constraints to which leaves are not subject, and this restricts their ability to concentrate only on photosynthesis. Nevertheless, research has shown that stem photosynthesis makes a significant contribution to both the growth and overall size of the plant. The overall contribution is dependent on several environmental factors but is especially important in plants that routinely lose their foliage to herbivores such as rabbits and caterpillars. Without photosynthetic stems, such plants would be forced to use only the stored energy in their roots to regenerate the parts lost to herbivory. For plants that are routinely damaged, this could eventually lead to death. For this reason, butterfly gardeners are especially appreciative of the fact that many larval plants are herbaceous.

Even woody plants start life with herbaceous stems. Wood is not produced immediately. At first, their stems are flexible and have an epidermis only one cell-layer thick. A young oak or pine tree, for example, has an herbaceous stem for at least its first few years, and during this time its green stem is supplementing the sapling's ability to photosynthesize and contributes to its overall growth rate. Young plants, whether they are woody or herbaceous, have a lot of environmental pressure to grow as rapidly as possible when they are young and most vulnerable.

Plants adapted to deserts and other extreme environmental conditions often rely extensively on their stems to provide for photosynthesis. Many have very small leaves or none at all. As leaves are the major source of water loss due to transpiration, reducing their size or eliminating them entirely is a trade-off that requires that photosynthesis be transferred solely to the stem cortex. Cacti are photosynthetic stems; their leaves have been converted to spines to further protect them from herbivory and the water loss that this would incur.

Young plants, even woody ones like this sapling camphor tree (*Cinnamomum camphora*), retain the ability to photosynthesize with their stems to maximize growth for the first several years.

Herbivory creates open wounds, and the wounds cause water loss. Plants have evolved ways to reduce this from occurring. The epidermis of many herbaceous and some woody stems provides protection by producing trichomes. These can be soft and fuzzy or stiff and spiny. Different plants, living in different environments, have evolved the type of trichome that best suits their needs. Some plants produce more than one kind of trichome, either in the same location or in different parts of their stems. Trichomes are very variable.

The thorns of a bramble thicket, made famous in many children's books, serve to protect herbivores, such as rabbits, from predators and to protect the plant itself from rabbit herbivory. Leaves are more easily replaced than stems. Brambles, such as raspberries and blackberries (*Rubus* spp.), take some precautions to protect their leaves, but they take extraordinary care to protect their stems. Of course, evolution serves the herbivore the same as it serves the plant, and various animals have evolved ways to feed on thorny plants. Giraffes, for example, have developed specialized lips in order to feed on the very spiny acacia tree (*Vachellia* spp.). Trichomes are not foolproof, but they limit herbivory to a small subset of potential herbivores.

The lack of wood in herbaceous plants makes it more difficult for them to provide the support needed to achieve the types of heights seen in many woody trees. All herbaceous stems achieve strength by maintaining optimal turgidity in their cells. When sufficient water is available, the fully turgid vacuoles press the cell membranes against the cell walls, and this keeps the stems upright. Such a mechanism is further enhanced by strands of pectin running through the cell walls and by maintaining a proper amount of calcium inside the cell wall itself. Plants with a calcium deficiency have difficulty maintaining their normal aspect.

Some herbaceous plants, however, develop stems that act like wood. They do this by reinforcing the cortex layer just beneath their dermis. Tall herbaceous plants, like sunflowers, use a system of collenchyma and/or sclerenchyma fiber bundles that run the length of the stem. These long, thickened fibers confer strength to the stem even when turgor pressure is reduced. Sunflowers can achieve mature heights much greater than other herbaceous plants because of this structure.

This extra strengthening reaches its greatest extreme in bamboos. Bamboos (there are more than five hundred species) are very specialized grasses. Some species can reach 40 feet tall or more. As monocots, bamboos do not

Bamboo is a specialized grass and therefore lacks wood. The strength of its stems is the result of specialized cells that are lignified.

produce wood, but their stems are used for a wide variety of products typically made from wood because their stems have great strength and durability. Bamboo has more compressive strength than wood, brick, or concrete and a tensile strength that rivals steel. Cells in the cortex and xylem of bamboo are lignified and contain additional silica, and their additional strength comes from this unique structural design.

Palms also are monocots. They do not produce wood either. They maintain strength in their trunks by adding new leaves as they grow upward. Each new leaf base wraps around the previous ones. Even as the old leaves are lost, their petioles remain. This overlapping arrangement produces sufficient strength to allow many palms to reach great heights. It also gives their trunks flexibility to withstand high winds. Wood cannot bend in this way.

Tree ferns do something similar to palms. Unlike most other ferns, their rhizomes grow upward, and the fronds are attached to the top instead of along the length. Embedded in their upright rhizome is a dense mantle of adventitious roots. As a tree fern ages and grows taller, these roots add girth and additional strength to the rhizome.

Woody stems are produced in a slightly different way than herbaceous ones. Woody stems begin life with a simple one-cell-thick epidermis, but this is replaced by a several-cell-thick periderm layer that develops beneath it as the plant grows. Periderm eventually completely replaces the original epidermis. It is a multilayered tissue system that contains the cork cambium that produces bark.

Much of the strength in woody plants is provided by the wood, not the periderm. The periderm is mostly bark, and bark is primarily present to protect the plant from outside forces such as fire, herbivory, and mechanical injury.

OUTER STEM STRUCTURE

All stems have the same basic structure. Leaves attach to stems at distinct locations termed nodes. Leaves are packed with chloroplasts, and these produce simple sugars during photosynthesis. They also require water to complete this process, and they lose water through transpiration. Respiration requires energy. Because of this, it is critical that the vascular system of the leaves are directly connected to the vascular system inside the

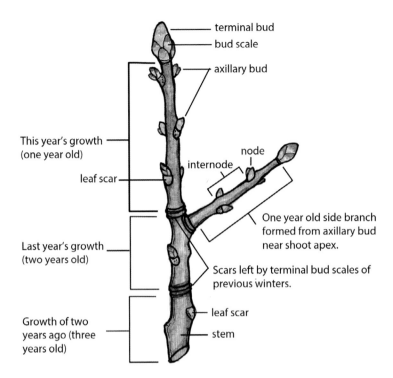

terminal bud
bud scale
axillary bud

This year's growth
(one year old)

node

internode

leaf scar

One year old side branch
formed from axillary bud
near shoot apex.

Last year's growth
(two years old)

Scars left by terminal bud scales of
previous winters.

leaf scar

stem

Growth of two
years ago (three
years old)

Basic stem structure.

stems. These connections are made at the nodes, regardless of whether the stem is herbaceous or woody.

The regions between the nodes are called internodes. There is nothing especially significant about them except that they provide most of the structure of the stem. Stems have to be able to "breathe," to exchange gases with the atmosphere. In herbaceous stems that are photosynthesizing, the epidermis of these internodes has stomata to allow carbon dioxide into the chloroplasts. Woody and herbaceous stems also need pores to release the carbon dioxide generated by respiration and to get oxygen to the living cells of the bark; this gas exchange is done using a different kind of pore called a lenticel. Lenticels are spongy-looking regions that come in a variety of shapes and colors. In fact, you could likely identify individual trees and shrubs from one another by closely examining the lenticels on the bark. Lenticels form around areas that were formerly stomata when the stem was herbaceous, but unlike stomata, they are not surrounded by guard and sub-sidiary cells. They are merely openings; they cannot open and close, but without them the living tissue just beneath the bark would suffocate and die.

Stems also contain terminal and axillary buds. Terminal buds are located at the ends of each stem and are the site where primary growth occurs. They may occur by themselves, or they may have several less dominant ones just beneath them.

In deciduous plants, the terminal buds are protected by thin, papery scales during the dormant period. These scales are modified leaves with no ability to photosynthesize and no succulence to make them attractive to herbivores. They remain closed up tight until the bud swells with the onset of a new growing season. The swelling forces the scales to drop away, and when they are gone a noticeable scar is left that encircles the stem just below the bud. Terminal buds are the only buds that leave this type of scar, and since this process is repeated each growing season, it is possible to measure the distance between each terminal scar and gain some insight into how much growth has occurred each season. There will be longer distances be-tween two scars when growing conditions were good than when conditions were poor.

Axillary buds are locations where new terminal growth could occur; most often at the site where a leaf is or was attached to the stem. If a stem branches, it will occur at the site of an axillary bud, and if a stem gets dam-aged, it will begin new growth at the site of the nearest axillary bud. Nor-mally, the portion of the stem between the damage and the nearest axillary bud will die and slough off. One major exception occurs when the main stem is damaged below any remaining axillary buds; the plant will not die but will instead produce a new bud near the damaged site before resuming growth. The vast majority of axillary buds never develop and grow. They are safety valves in case they are needed, but the plant's hormonal system keeps them from developing further.

INNER STEM STRUCTURE

In most ways, the inner structure of a stem resembles that of a root. The major differences lie mostly in the way the three tissue systems are designed. Stems of all vascular plants have dermal, vascular, and ground tissue.

The dermal layer of herbaceous stems and the periderm of woody ones have only minor differences from the dermal layer of roots. Although the cell types are virtually identical, the root dermal layer produces trichomes (root hairs), whereas the stem may not. When present, stem trichomes take a completely different form, and they can assume a great many shapes. None is involved in absorbing water. The dermal layer in many herbaceous species also produces a waxy cuticle on the surface of the stem to impede water loss. Such a function is incomprehensible for a root. Roots are designed to maximize water absorption, not impede it.

Stems transport water from the roots to the top of the plant and transport other materials, such as sugars and hormones, in both directions between the foliage and the roots. These materials are carried within the xylem and phloem, respectively. The cells of the vascular tissue in a stem are identical

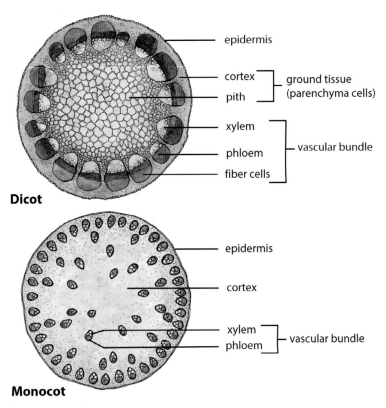

Stem cross section.

in structure to those of a root. The cells are arranged a bit differently in a stem, but that does not affect the way they work.

The ground tissue of a stem is also largely composed of parenchyma cells, but because stems are the primary organ for providing structure, they differ in the arrangement of collenchyma and sclerenchyma fiber bundles. Collenchyma and sclerenchyma fibers are present in the stem cortex just inside of the dermal layer and surrounding the vascular bundles. This also is true for young woody plants as they are herbaceous as juveniles. The "strings" in celery stalks are collenchyma, and they lie just below the dermal layer to provide support. The fibers we use to make linen and rope are sclerenchyma, and they occur in much the same location.

Collenchyma and sclerenchyma fibers also surround the vascular bundles of herbaceous and woody plant stems to provide support here as well. The vascular bundles in monocots are scattered throughout the stem; therefore the reinforcing fibers are scattered too. In other flowering plants and in gymnosperms, the vascular bundles encircle the stem just beneath the dermal layer. This puts the collenchyma and sclerenchyma fibers in this same location, giving the young stem added strength but allowing it to bend.

Stems have to remain strong to support the various functions of the plant. The position of the leaves in relation to the sun affects photosynthesis, and that is largely determined by the ability of the stem to maintain that configuration. The height of the plant also influences the amount of sun a plant receives, a relationship determined by the stem. Most stems quickly alter their palatability as they mature in an effort to do their job most effectively. If you've ever grown asparagus (*Asparagus officinalis*), for example, you've encountered that. The succulent stems we wish to harvest and eat, quickly become woody if not cut in just the first few days. If you are a butterfly gardener, you've likely noticed that female butterflies always lay their eggs on the newest growth possible and that caterpillars avoid the older stems until there is nothing left to eat. Not all stems are involved in photosynthesis, but all are intimately involved with providing structure.

SPECIALIZED STEMS

Stems come in a wide variety of shapes and styles. Most above-ground stems are recognizable to us as stems, but others are less obvious. By definition, stems are the plant organ that connects the leaves, needles, or fronds to the roots. Roots are the organ that draws water and nutrients into the plant. Many things that we know as "roots" are actually stems.

By definition, tubers are stems. Potatoes, yams, Jerusalem artichokes (*Helianthus tuberosus*), dahlias (*Dahlis hortensis*), and the like are not roots. Carrots, radishes, and beets are. The latter are involved with pulling water out of the soil in addition to storing extra starches for the plant's future needs. Tubers have only a storage function. If you keep a potato too long on the shelf, it will begin producing new stems with leaves and roots from each eye. The energy for that growth comes from the tuber. Over time, the original potato will shrivel as its stored energy is transferred to the new plant growing from that eye. I once kept an air potato (*Dioscorea bulbifera*) in my windowless office to show my plant biology students. The vining stem produced by this tuber grew for more than a year, living solely on the energy stored in the tuber, before it expired. At that time, the original tuber had almost disappeared. Tubers are specialized storage stems.

Ginger "root" (*Zingiber officinale*) is a stem, not a root. Ginger and a great many other things we call roots are really rhizomes. That includes a variety of spices used mostly in Asian cooking—turmeric (*Curcuma longa*) and galangal (*Alpinia galangal*) as well as the underground stems of horticulturally significant plants such as iris (*Iris* spp.), canna (*Canna* spp.), and bamboo. Some orchids develop rhizomes, and all ferns have them. Technically, a rhizome is an underground stem or one that creeps affixed to the growing substrate. They produce new shoots from the upper surface and roots from below.

Rhizomes develop from axillary buds and grow perpendicular to the force of gravity. If a rhizome is separated into pieces, each piece can develop a new plant if it contains an axillary bud. Rhizomes store starches, proteins, and other nutrients and use these when new shoots are formed or when the plant dies back for the winter. Plants that develop rhizomes spread laterally throughout the landscape, and some can become pests because of this.

If the lateral stems run above the ground, or just below the surface, and form new plants sporadically at nodes along its length, they are called stolons. Strawberries (*Fragaria* spp.) and the common house plant known as a "spider plant" (*Chlorophytum comosum*) produce stolons. There is some disagreement among botanists as to what exactly separates a stolon from a rhizome, but a stolon is never the main stem of a plant; it sprouts from an

Plant stems are often specialized to serve specialized purposes. Vining stems have evolved in many different types of plants to enable them to climb into the canopy of forests and reach sunlight.

existing stem and generates new shoots at the end. Stolons arise from the base of the plant. Unlike rhizomes, they also generally die once the new plant is generated at the terminus.

Bulbs are modified leaves with a hardened patch of stem beneath. Onions (*Allium cepa*), garlic (*Allium sativum*), leeks (*Allium ampeloprasum*), and a great many other culinary herbs are bulbs, not roots. So are many of our favorite landscape flowers such as tulips (*Tulipa* spp.), daffodils (*Narcissus pseudonarcissus*), and jonquils (*Narcissus jonquilla*). This arrangement is easy to see in a leek, but it becomes obvious in an onion too if you keep it too long in the pantry. Much of the bulb soon becomes mushy as leaves start to emerge from the top. The reason why restaurants can turn an onion bulb into a deep-fried "flower" is because each "petal" is really an undeveloped leaf awaiting its turn to finally emerge.

The stem of a bulb is actually the base where the roots develop. In an onion, this is a small hardened patch that we normally cut off prior to slicing it. Bulbs are extremely good at storing energy over extended periods while the plant is dormant. This is why we can dig them up and store them over winter for spring planting without damaging them. The stem portion generates new roots when growing conditions are right, and new leaves will develop and emerge from the collection of leaves that comprise the bulk of the bulb. Many bulb-producing plants reproduce asexually by developing small bulbs off the base of the stem portion. These so-called seed bulbs allow new plants to develop from established bulbs even if the older plant dies.

Some plants develop underground bulb-like structures that are solid and not composed of modified leaves. Such structures are known as corms. Blazing stars (*Liatris* spp.), taro (*Colocasia esculenta*), crocus (*Crocus sativus*), and gladioli (*Gladiolus* spp.) are commonly propagated plants that grow from a corm. Corms are solid, and if you were to cut one, it would not be arranged like an onion with concentric circles. Corms produce adventitious roots, and many also have a second type of specialized roots, known as a contractile roots, that serve to pull the corm back beneath the soil surface if environmental forces expose it. At the top of the corm, one to several buds grow into shoots that produce normal leaves and flowers. Over time, small corms, or cormels, may form at the base of the original corm. These serve the same purpose as seed bulbs at the base of a bulb.

Some tendrils are modified stems. Those that grow from the tips of leaves, like peas (*Pisium sativum*) and pole green beans are modified leaves, but others arise from the stem. Examples of plants with stem tendrils are cucumbers (*Cucumis sativus*), grapes (*Vitis* spp.), and passionvine (*Passiflora* spp.). Tendrils are specialized structures that attach to neighboring plants or structures and provide the support needed to help pull the main stem higher into the vegetated canopy. Tendrils by themselves are not especially

Tendrils, like in this Virginia creeper (*Parthenocissus quinquefolia*), are especially well adapted to help it climb.

strong, but their strength is greatly enhanced when they occur in great numbers. The coiled tendrils also absorb and dissipate energy when the plant is subjected to strong wind.

Tendrils are one of the best examples of thigmotropism, a plant's ability to react to touch. As it is in animals, a sense of touch is critical to plants, and it directs many of their responses to the world around them. Plants can detect touch with all of their organs, but it is especially well developed in its stems.

Vines without tendrils simply climb by winding around the stems and branches of neighboring plants. Wrapping the stem without the use of specialized roots or tendrils requires a well-developed thigmotropism. Wisteria (*Wisteria* spp.) is a widely grown ornamental vine that climbs in this manner.

Plants that are not vines still need to sense and react to the touch of neighboring plants. It is not uncommon for the trunks of adjacent trees to nearly merge into one another over time to form a reinforced single base. Sensing one's neighbors is also important in reducing conflict. In a forest, the branches within the crowns of adjacent trees rarely touch, a phenomenon known as "crown shyness." The exact physiological basis of this is not certain, but the variety of hypotheses and experimental results suggests that there are multiple mechanisms involved.

Trees, because of their height, also need to sense their position in the world around them and make decisions regarding their configuration. One way trees do this is to shed branches in a way that maintains their equilibrium. Trees also can regain their sense of balance following a major storm event that partially dislodges them. A tree left severely leaning after such a storm often directs the growth of its remaining branches in a way that restores its center of gravity. How this is accomplished specifically is not known, but it is clearly a result of its sensory awareness and worthy of future studies.

7

Leaves

Leaves are the plant organ most responsible for photosynthesis and often are what gives a plant its distinctive and aesthetic character. Most gardeners and naturalists can identify a plant by its leaves alone, although few can do the same by its stems. We associate certain leaf patterns and shapes with individual plant species and we diagnose a plant's overall health based on their condition. Leaves become the manifestation of our gardening prowess, and few gardeners, other than those interested in butterflies and

Leaf shape is extremely varied in the plant world and can often be all one needs to identify each species.

moths, fail to stress if the foliage of their landscape plants starts to look ragged from the rest of the world chewing on it.

LEAF STRUCTURE

Leaves come in all shapes and sizes, but their basic structure is very similar. Leaves basically have two parts, the leafy part, which is correctly called the blade, and the petiole, the stem that attaches the leaf to the branch.

Leaf blades can be solitary or compound. A solitary leaf has a single blade attached to the petiole, whereas a compound leaf is composed of two or more blades (leaflets). In some plants, each individual leaflet is composed of more than one leaflet. Such leaves are twice-compound. Many legumes (bean and pea family plants) have compound leaves, but having simple or compound leaves is not necessarily related to taxonomy. Some leaves, like those of most grasses and other monocots, do not have a petiole; the leaf blade wraps around the main stem of the plant and attaches to it. These types of leaves are called sessile.

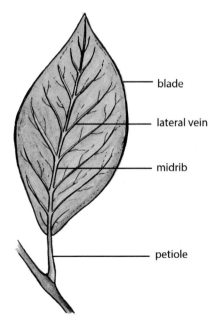

blade

lateral vein

midrib

petiole

Basic leaf structure.

Sugars produced in leaves during photosynthesis travel through the petiole to reach the rest of the plant, and water and nutrients pulled up by the roots use it to reach the leaf chloroplasts to supplement photosynthesis. Plants without petioles do this at the point where the leaf blade directly attaches to the main plant stem.

No leaf lives forever. When leaves die, they separate from the main plant stem at this point also. Shedding old leaves may seem a simple process, but it requires preparation by the plant. When a leaf falls off the main stem, the channels of xylem and phloem that were once connected must be closed off. To keep these open would invite pathogens to enter the plant and result in water loss. The process is quite analogous to what occurs in animals. Open wounds need to be healed as quickly as possible. Animals do this by forming a scab after the wound has occurred. Plants form this barrier prior to leaf fall. They anticipate normal leaf fall and create a barrier. When leaves are forcibly removed, this process is initiated immediately.

When a plant prepares to lose its leaves, it readies the site ahead of time so that an open wound does not result. These leaf scars are evident well after the leaf has fallen.

When leaves become senescent or are forcibly removed, plants produce hormones that cause a barrier to begin forming between the petiole and its point of attachment on the stem. The xylem and phloem channels are closed off and plugged with a highly waterproof substance known as suberin. All of this leaves a permanent leaf scar that is visible on close examination.

Pulling leaves off plants (or pruning them) creates avenues for airborne and waterborne pathogens to enter. When plants can't prepare ahead, they have to seal their wounds after the fact. They do this rather quickly and, most of the time, accomplish it before anything serious happens. Problems can occur, however, just as in animals. Plants are more likely to suffer repercussions from damage and pruning if they are already stressed or if their growing conditions are less than ideal. A hospital will not operate on you if your vital signs have not been stabilized, and they won't perform the operation in a back alley filled with trash using an unsanitary blade. If you must prune your plants, do not treat them any differently. Perform the operation when your plants are healthy and growing conditions are optimal, and use clean, sharp pruning shears.

The leaf blade is comprised of all three tissue types, and each provides a significant function. In most plants, the dermal layer on the upper surface is arranged differently than the lower one, though both are one cell thick. As the upper surface of most leaves is held at an angle that maximizes its reception of incoming sunlight, it also is the surface most susceptible to water loss. Therefore, the upper dermal layer often lacks stomata. Water lilies must have their stomata on the upper surface because their lower surface is just below the waterline. The upper surface is the only one that can let in the carbon dioxide required for photosynthesis. Conifers and other needle-leaved plants put most of their stomata on the side of the needle most distant from the dominant direction of the sun, and they sink all of them beneath the surface to further reduce the potential for water loss.

The dermal layer of a leaf may produce a waxy cuticle. Cuticles are generally produced on the upper leaf surface. They are clear so that sunlight can pass through to the chloroplasts virtually unimpeded, but their waxiness significantly reduces water loss from the cell layers just below. Most plants adapted to high levels of sunlight, such as those in the tropics and in deserts, have leaves with noticeably shiny cuticles.

The leaf dermal layer may also produce trichomes. Trichomes come in all sizes and shapes, but they are most often "hairy" or spiny. Thick felt-like layers of trichomes are generally produced on the underside of leaves and not the upper side. Trichomes on the upper surface would block sunlight from reaching the chloroplasts. Only plants adapted to dry and extremely high-sunlight habitats, such as deserts, produce dense coverings of trichomes on their upper leaf surfaces. Trichomes on the upper surface are normally finer, more scattered, and often used in defense. The long spines on the upper surface of tropical soda apple (*Solanum viarum*) leaves protect them from herbivory but don't affect photosynthesis. The same is true for the stinging hairs of nettle (*Urtica dioica*) and the irritating hairs of poison ivy.

Leaf trichomes come in all styles and shapes. Some, like this dense mat of "hairs" found on the underside of southern magnolia (*Magnolia grandiflora*), provide much more than aesthetic interest.

Photosynthesis occurs in the middle layer of leaves, the region known as the mesophyll. A lot happens here. The mesophyll contains the chloroplasts, the cells that contain the chlorophylls, as well as the chromoplasts that contain the carotenoid pigments. Sunlight has to reach the mesophyll for photosynthesis to work. The mesophyll is the work horse of the leaf.

Each mesophyll cell contains a great number of chloroplasts. Though this varies by plant species, studies have determined that each cell can contain more than two hundred chloroplasts. Depending on the size of the leaf, the numbers of chloroplasts involved in photosynthesis at any one time is staggeringly large.

Mesophyll is composed solely of parenchyma cells, but different types of plants have their mesophyll arranged a bit differently. Some species have irregular, or even branched, parenchyma cells (Y-shaped). Ferns, gymnosperms, most monocots, and a few other flowering plants are structured this way. The majority of flowering plants, however, have two distinct layers of parenchyma mesophyll cells—palisades and spongy mesophyll.

Palisade parenchyma is usually directly beneath the epidermis of the upper leaf surface, and the spongy parenchyma fills the space beneath it. Palisade parenchyma is a series of tightly packed columnar-shaped cells, while spongy parenchyma cells are irregularly shaped with significant air spaces between them. These air spaces are connected to the atmosphere by the open stomata on the underside of each leaf. Both types of parenchyma contain chloroplasts, but each remains specialized to its primary function. The tightly packed columns of palisade parenchyma at the upper surface of the leaf receive the most sunlight and perform most of the work needed in the initial phase of photosynthesis. The spongy parenchyma is designed to maximize the amount of carbon dioxide inside the mesophyll in order to complete the process and maximize growth. Such an intricate balance of shared responsibilities demonstrates the wisdom of evolution.

Inside each fern frond, pine needle, and leaf is vascular tissue designed to collect the glucose produced by photosynthesis and transport it to other regions of the plant as well as transport the water required for this and for transpiration. This vascular tissue is arranged in bundles of xylem and phloem and often surrounded by a sheath composed of collenchyma and/or sclerenchyma cells for added strength and parenchyma cells to store starches. Certain types of tropical plants—those referred to as "C4"

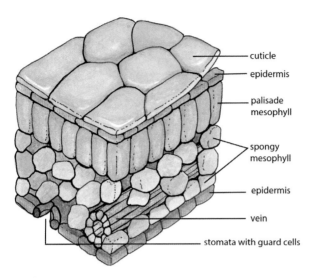

cuticle
epidermis
palisade mesophyll
spongy mesophyll
epidermis
vein
stomata with guard cells

Leaf cross section.

plants—contain parenchyma cells with chloroplasts in their vascular bundle sheaths.

Just like in our own vascular system, where larger arteries and veins end up as tiny capillaries that bathe our cells in oxygenated blood or collect it to send it back to the heart, a plant's vascular bundles must do something similar. Leaf veins branch successively off the main vascular bundle(s) that connect to the main stem until they are just a few cells away from any parenchyma mesophyll involved with photosynthesis. The products of photosynthesis are then packed into the phloem cells for further transport, and the water and nutrients are released out the ends of the xylem tubes to bathe the cells around it.

The arrangement of the vascular bundles in leaves differs between various groups of plants and is easily observed when looking at them. Veins typically branch in a somewhat netted pattern in fern fronds and in most flowering plants while they run parallel to one another in gymnosperms and in monocots. Both systems work efficiently and most likely have evolved more for the strength they provide the leaf than for any hypothetical differing need that might arise for the transport of materials. Leaf veins do more than move the products carried inside the xylem and phloem; they also provide the structural support needed to keep the leaf at its optimal configuration to the sun. Large leaves of all venation styles have a large central midrib vein for this purpose.

The needles of some conifers have one other type of transport tube known as resin ducts. The resins produced by pines and other conifers formed the foundation of the "naval stores" industry, producing the turpentine, pitch, tar, and rosin required by the navies of developed nations during the colonial period and beyond. These resins provide us waterproofing products as well as products that resist decay. In the living tree, they do the same as well as provide the services of an antifreeze. Conifers manufacture these resins and transport them throughout the plant, even to the tips of the needles. A pair of resin ducts, one to the outer side of each vascular bundle, is present in each needle. Among other things, these resins allow various conifers to keep their foliage intact through the most frigid winter temperatures.

GROWTH AND DEVELOPMENT

Leaves are evident in the earliest stages of a plant embryo's life. The shoot apical meristem is where the early stem begins development, but it has two roles: it is the source of the new cells that are needed for stem growth; and it is the site of small cellular outgrowths, called leaf primordia, which develop into the leaves. The leaf primordia occur in predictable positions along the stem, being evenly spaced around the circumference of the stem apical meristem. Their future development into leaf buds is triggered by a specific protein, known as expansin, coupled with the plant's production of the hormone auxin.

Embryonic leaves become evident before the plant germinates, and the production of successive leaf primordia after germination occurs regularly along the stem regardless of environmental factors. Recent research has shown a strong correlation between the number of initiated embryonic leaves and the number of expanded leaves once the plant is actively growing, indicating that the production of new leaves capable of assisting in photosynthesis is the primary trigger telling the growing plant to make more. This was true regardless of environmental factors such as temperature and soil moisture.

Embryonic leaves, known as "leaf primordia," are evident very early in the development of the plant embryo. Photograph by Mira Janjus.

The shape and surface area of a leaf are largely determined by the plant's genetics, but environmental conditions can influence these factors to some extent. Research has shown that leaf size and shape are altered by water availability and light. Nutrient-poor environments also influence the overall size of leaves. The most significant effects were seen in simple leaves rather than compound ones. Compound leaves seem to be more efficient at transferring heat than simple leaves and may allow plants in harsh environments to maintain a higher leaf surface area than would be predictable if the plant had simple leaves.

In low light, there are two major forces somewhat at odds with one another. Low light reduces growth and makes it more difficult to generate leaves, but producing leaves with a greater surface area increases the ability of each leaf to produce more energy. Plants attempt to balance those two forces as much as their underlying genetics allow. Light seems to affect leaf growth more than any other environmental variable, especially in simple leaves. In reduced light, simple leaves tend to elongate and become broader at the base near the petiole. By this means, leaves can maximize their light-

Plants in a forest compete for light through the positioning of their leaves.

capturing surface area but also reduce the chance that they will shade the leaves below them. It is an ingenious strategy.

Leaves also may compete with one another for light when it is limited. We sometimes think of plants as passive beings, but they are no different than animals when it comes to competing with one another for resources. Recent research suggests that plants restrict this competition when growing next to siblings, but exhibit no such inhibitions when grown with unrelated specimens of the same species or with different species altogether. Unrelated plants grown in close proximity direct their growth into leaves and away from reproductive growth (e.g., flowers and fruits) and long-term energy storage (e.g., roots and tubers). In other words, competing plants make a conscious decision to sacrifice their long-term success for their short-term gain. It is more important to win the all-important battle for light in the present so as not to lose the ability to survive in the future. This is especially evident in plants growing within three feet of one another. Leaf competition is more related to gaining a superior height in the canopy than in outcompeting one another in terms of the number and size of leaves each produces. Plants in competition with one another put their energy into gaining a height advantage, but they also put more energy into producing more leaves near their crowns and less into leaves below it. Then they have to worry about maintaining that advantage. Plants in a competitive relationship with their neighbors can't relax any more than animals can.

LEAF MOVEMENTS

Plants also move their leaves to gain advantages or to reduce their exposure to risks in the environment. Not all leaf movements are the same. Like animals, plants have biological clocks that allow them to respond to changes in time. Among the more obvious clock responses are "sleep movements" such as the closing of flowers and changes in leaf position that many plants display at night. As leaves are the plant's most important organ for photosynthesis and also one of its most succulent to herbivores, it makes great sense to shift the position of its leaves or fold them up completely during the evening hours when the sunlight needed for photosynthesis is absent and a great many herbivores, like rabbits and rodents, are most actively feeding.

Many plants fold their leaves up at night, when they are not needed for photosynthesis. This type of "sleep movement" likely makes them less noticeable to nocturnal herbivores.

These types of movements are under control of the plant's biological clock, and this clock requires environmental cues to stay synchronized throughout the twenty-four-hour day. Light and temperature are two of the most important environmental stimuli since they usually change between night and day. Biological clocks in plants, as well as animals, persist for days even in the absence of a daily change in environmental cues; however, in most plants the response will quickly start to drift and need to be reset by an appropriate external stimulus. Sleep movements, based on a plant's biological clock, allow a plant to reorient its leaves in unison with the timing of sunrise. Plants make accurate judgments each day about when sunrise will occur, and they position their leaves and/or flowers in unison with it.

Plant movements that occur in response to environmental stimuli are called nastic movements. Unlike the thigmotropic movements of stems and tendrils, the direction of a nastic response is not dependent on the direction of the stimulus. Some of the most spectacular plant movements are nastic movements. These include the closing of the carnivorous Venus flytrap leaf when it captures prey or the folding of the sensitive mimosa (*Mimosa pudica*) leaf when it is disturbed.

Nastic responses tend to be quicker than those of thigmotropic movements. They are not based on the plant's growth and cell division as is the coiling of a vine; they depend on pre-accumulated turgor mechanisms in-

Some plants, like this sensitive pea (*Chamaecrista nictitans*), have a specialized structure, a pulvinus, at the base of their leaves that is sensitive to touch and closes the leaf.

side existing cells. Plant physiologists have identified specific signaling molecules that such plants release, called turgorins, which mediate the loss of turgor. In species with the fastest response time, like the Venus flytrap, the plant's vacuoles provide temporary, high-speed storage for calcium ions.

Many so-called sensitive plants have specialized structures at the base of their leaves called pulvinae. A pulvinus causes the leaves to rapidly fold when touched, and this seemingly protects other leaves from potential herbivory by making them appear less succulent. The pulvinus is a motor structure consisting of a rod of sclerenchyma cells surrounded by collenchyma. The whole leaf-folding process is dependent on an electrical impulse being passed down the leaf, not unlike the electrical stimuli sent down from the tips of our fingers to our brain. In plants, the impulse is stopped at the pulvinus, where the petiole intersects with the stem. This ensures that only those leaves that are physically touched will fold and that the others can continue their role in photosynthesis.

Not all nastic movements are related to leaf closure. Some species of asters, for example, use it to facilitate pollination. When an insect lands on their flower, the anthers shrink and rebound, loading the insect with pollen and sending it off again, somewhat like a trampoline. The effect results from turgor changes in specialized, highly elastic cell walls of the anthers.

DIFFERENT KINDS OF LEAVES

Some plants make specialized leaves for purposes other than photosynthesis. It is not uncommon, for example, to produce leaves at certain times of the year to assist reproduction. Most ferns produce their spores on the underside of their mature fronds, but a few do so only on specialized fronds that have almost no photosynthetic role and are present just long enough to form and release the spores. In the region of Florida where I reside, the most common species like this are netted chain fern (*Woodwardia aerolata*) and the osmunda ferns (*Osmunda* spp.). This strategy allows these ferns to maximize their reproductive investment in just a few fronds for a very limited time and use the rest of their fronds year-round for photosynthesis.

Some of our most favorite flowering plants are not really known for their flowers but for the leafy bracts that surround them. The brightly colored bracts of our holiday poinsettias are leaves, and they surround a collective

Ferns reproduce by specialized structures known as "sori," often found on the undersides of the leaves (fronds).

center of small green flowers that would go largely unnoticed otherwise. In fact, a great many of the euphorbias that we grow as house and landscape plants have similar bracts surrounding innocuous greenish flowers. The bright-white bracts of flowering dogwood (*Cornus florida*) provide the same function. Flowers are an energy-rich investment on the part of the plant. If that function can be provided by a leaf instead of a petal, it reduces that investment. Plants like those listed above can produce dozens of flowers and surround them for the entire time they remain open, with the investment of only three to five modified leaves.

Some flowering plants, mostly dicots, produce small accessory leaves at the base of the petiole of the main leaf or along the stem. These leaves are called stipules. Most stipules come in pairs, though a few actually encircle the stem and a few are solitary. In some plants, the stipules die, shrivel, and are discarded relatively soon after the leaf has finished it growth. Such stipules leaf a stipular scar just as fallen leaves leave a leaf scar. Some stipules look "leafy," but many others act as thorns and spines.

The original function of stipules is obscure, but it is likely they were mostly involved as protection for the emerging leaves. Conspicuous stipules, covering the buds, can be observed in a great many plants. In figs (*Ficus* spp.), relatives of teak (Family Dipterocarpaceae), and red mangroves (*Rhizophora mangle*), the stipule is well developed and appears to serve in bud protection; however, the majority of flowering plants show no disadvantage by lacking them, and in many the stipule is small and vestigial, without any obvious

Some plants produce accessory leaves, known as "leafy stipules," adjacent to their primary leaves. These likely assist with photosynthesis.

function. Leafy green stipules are likely photosynthetic, but little research has been done on this, and their importance to the plant's overall growth and development is poorly understood.

Some stipules become hardened with lignin, the compound that also hardens wood. Such stipules mostly function as spines. Stipular spines may be solitary or occur in pairs at each node along the stem. Some, like in the bullhorn acacia (*Vachellia comigera*) of Mexico and Central America, have developed hollow thorns and formed mutualistic relationships with ants, which can then defend the plant from herbivores. Stiff stipular spines are also significant in protecting many Old World succulents, like the commonly grown crown-of-thorns (*Euphorbia milii*), from herbivorous mammals. Some of these even develop specialized nectar-producing structures, called nectaries, which provide a benefit to the ants and keep them nearby. Many New World desert and chaparral species, like *Ceanothuus* spp., have corky stipular spines. Cactus thorns are not modified leaves; they are produced by buds along the stem and therefore are modified stems. The same is true for the thorns on roses and their relatives (Family Rosaceae).

Most tendrils, as discussed above, are modified stems, but a few are modified leaves. Such tendrils are produced at the leaf tip and not along the stem. Climbing peas and beans (Family Fabaceae) produce this type of tendril. The greenbriers (*Smilax* spp.) also produce stipules (i.e., leaves) that are modified as tendrils.

Perhaps the most specialized leaves are those found in carnivorous plants. The traps of the Venus flytrap are modified leaves, but so are the sticky leaf pads of sundews (*Drosera* spp.), the bladder-like trigger traps of bladderworts (*Utricularia* spp.), and the funnel-form traps of the New World pitcher plants (*Sarracenia* spp.) and Old World pitcher plants (*Nepenthes* spp.). Such modifications to both trap and digest animals remain perhaps the most amazing plant adaptation to date among a host of other amazing adaptations.

Carnivory occurs in areas where soils are low in nitrogen and other essential elements. Animal life supplements nutrient deficiencies and allows the plants to survive and grow at a normal rate. Most carnivorous plants make their own digestive enzymes; others depend on bacteria to produce them. The bacteria cause the captured prey to rot, and the plant absorbs the nutrients. Some plants use a different method that is more unappetizing. They use bugs and insects as helpers. Assassin bugs crawl around certain types of sundews, for example, and eat the insects that have been trapped. Then these bugs poop, and their feces provide the nutrients the plant requires.

Leaves are not a requirement, and some plants have lost them or reduced their function to very specialized purposes. Most cacti, for example, have shed their leaves and developed photosynthetic stems. The loss

The leaves (and stems) of this Indian pipe (*Monotropa uniflora*) lack chlorophyll altogether.

of leaves for photosynthetic purposes reduces their risk of being fed on by herbivores. Other types of plants, like certain orchids, are leafless and are able to photosynthesize with their roots. Still others don't need to photosynthesize at all because they are root or stem parasites. Mistletoe may be the most famous plant parasite, but they maintain photosynthetic leaves. Some, like the Indian pipe/ghost plant (*Monotropa uniflora*), lack chlorophyll altogether; their rudimentary leaves are white and "ghostly."

8

Reproduction

If plants didn't reproduce, there would soon be no more of them. It's a simple fact of life. For gardeners, however, it goes much deeper than that. Few of us grow plants only for their foliage. Without the cones, flowers, and fruit, most of the plants we add to our landscapes would not provide the aesthetic and practical functions for which we've chosen them. We don't just expect reproduction to happen; we count on it. Understanding how it works in plants is important.

ASEXUAL REPRODUCTION

It is not necessarily true that there wouldn't be baby plants without sex. Asexual reproduction in animals is relatively rare, especially once you reach the rung on the evolutionary ladder where animals have developed backbones. A great many plants, however, retain their ability to produce new plants without the fuss and bother of sexual intercourse. Asexual reproduction bypasses the great energy investment required to produce reproductive structures, join gametes, and nurture offspring until they are capable of independent living. The downside to asexual reproduction is that all of your offspring are identical to you. Sexual reproduction ensures genetic diversity, and this is almost always a better strategy in nature.

Some plants produce new individuals by fragmentation (breaking apart into separate pieces) with the result that all parts are capable of starting growth and living independently. This is especially true for bryophytes (mosses, liverworts, and hornworts), but many higher plants are capable of it, depending on where the break occurs. Most plants that grow from an underground rhizome (ferns and some flowering plants) routinely form new independent plants by having their rhizome broken apart. Flowering plants that develop from bulbs and corms do the same. Bulblets and cormels are identical to their parents. Other plants retain the ability to make new plants from their stems. The stolons of a strawberry, for example, give rise to dozens of new plants as they creep across the soil surface and touch moist soil at their nodes.

Certain plants come equipped to form new plants from their stems by having root primordia already present. We see these as bumpy nodules on the stems of tomatoes, coleus (*Plectranthus scutellarioides*), and pothos (*Epipremnum aureum*). If these nodules touch the soil under the right conditions, they will form fully functional roots. Many willows (*Salix* spp.) have leafy bud-like root primordia on their stems for the same reason.

Many plants, such as this pipevine (*Aristolochia* spp.), retain the ability to quickly produce roots from their stems.

We routinely use the capability of plant stems to form roots in horticulture by propagating new plants from cuttings. In the simplest of plants, the only thing necessary is to put the cutting in water and wait for roots to develop at the cut end of the stem. Even certain woody trees, like the tropical gumbo limbo (*Bursera simaruba*), can root without additional help if their cut branches are quickly put in moist soil. In other types of plants, cuttings will root only if first treated with a root-stimulating hormone. The cut end develops a callus of suberin that plugs the wound and prevents the further loss of sap. It also protects it from pathogens. The living tissue above this callus can then form adventitious roots. Four stages need to occur for roots to form from a cutting, regardless of the type of plant. The cells behind the callus have to dedifferentiate; cells with existing specific functions have to become nonspecific. These cells then need to form root initials; the dedifferentiated cells near the vascular bundles need to become meristematic and become capable of producing new cells. Root-initiating cells in this area then form root primordia, and eventually the primordia form new functioning roots from the cutting and become connected to the vascular tissue of the original stem. The plant hormones in the root-stimulating compounds available in all gardening stores facilitate this process or stimulate it to happen in certain types of cuttings that would not form roots on their own.

Some plants can also root from their leaves, either from the base of the petiole or the margin of the leaf blade. Some plants do this naturally while others can be induced to form roots and then new plants by allowing the cut to form a callus and then supplying them with a mixture of various hormones. The rise of tissue culture in mass-producing clones of particularly interesting specimens of a plant species is all based on this intrinsic ability of plants to reproduce asexually.

A few plants reproduce asexually within their flowers by a process known as apomixis. Plant biologists recognize a great many types of apomixis, but for the purposes of this book it is not necessary to draw fine lines between them. Certain flowering plants do not produce seed by pollination and the ultimate fertilization of the ovules. Their seeds develop without pollen through a somewhat complicated process. The common lawn weed dandelion (*Taraxacum officinale*) normally produces seed through apomixis. Their flower heads attract pollinators, but their activity is not what produces

Meadow garlic (*Allium canadense*) produces flowers and reproduces like other flowering plants, but it also reproduces new
plants asexually by producing bulblets.

the ripe, fluffy seed. The seed would form regardless. Other apomictic plants
produce bulblets atop their flowering stems instead of flowers. Meadow
garlic (*Allium canadense*) and its relatives commonly reproduce this way.
Plants produced by apomixis are genetically identical to their parent. In
other words, they are clones.

SEXUAL REPRODUCTION

Sex guarantees genetic diversity, and this allows a plant to have offspring with differing abilities to meet a changing world. Even if you are the ultimate specimen to succeed in today's world, there is nothing to guarantee that tomorrow will be identical to today. In fact, the history of the world almost ensures that things change and that the intelligent choice is to be prepared for it.

In a stable world, plants may opt for asexual means to produce their offspring, but even then they are likely to put some of their energy into sex. Sex is always the more complicated route. It takes energy to get ready to find a mate, and it takes even more to locate and enlist one to make children with. This energy investment is nothing compared to what it takes to nurture the developing embryos until they are mature enough to fend for themselves. Plants are really no different than animals in these regards. The sex life of a plant is far more active and directed than most of us think.

It can take a lot of preparation for a plant to get ready for sex. For one, they must build up sufficient energy reserves to be successful. Plants make choices about what to do with the light energy they receive, and having sex is low on the list and comes well below meeting basic metabolic needs and growth. Sex can always come later. Some plants take years to build up sufficient energy reserves to reproduce, and then, after they reproduce, they die. In a way, these are really long-lived annuals. Trying to save up enough energy for sex more than once in their lives is simply overwhelming given the conditions under which they are growing. Some of the best examples are the agaves (*Agave* spp.), the so-called century plants. As their name implies, they often take decades to mature enough to produce their spectacular flower spike and then set seed. After that, they die, regardless of the growing conditions.

Some individual plants die after they reproduce simply because they can't recover the energy they've just expended. It is not uncommon for this to happen when a plant is growing under less than ideal conditions. Though most plants won't produce reproductive structures if they are energy-stressed, many will do so when faced with death. They put whatever energy they have left for one last attempt to leave offspring behind before they die. The drive to produce progeny is just as strong in plants as it is in animals, but sex is energetically taxing. If a plant is not vigorous but produces a flower stalk anyway, it may be best to cut it off when it is first emerging. By preventing the plant from using its energy reserves for reproduction, its energy can be used for future growth and increased vigor.

Plants also have to reach a certain level of maturity before they can become sexually active. What is true for animals is equally true for plants. Different types of plants reach sexual maturity at different ages. Some are

Plants, like animals, have to reach sexual maturity before they can reproduce. Some, like oaks (*Quercus* spp.), may take several decades to reach this stage.

precocial and become parents at a very early age, whereas others wait for decades. Some of this is influenced by their growing environment (same as animals), but a lot has to do with their genetics and their life history that has been shaped by their ecology. Generally, plant species with the greatest potential longevity wait the longest before initiating sexual reproduction. Many oaks (*Quercus* spp.) do not begin producing acorns until they are at least twenty-five years old, but it can sometimes take fifty years to reach this stage. In our thirty-year-old neighborhood, none of the live oaks have begun producing acorns, but all of the laurel oaks (*Q. laurifolia*) have. Both age and growing conditions will influence the basic genetics of how quickly a plant will produce sexual structures like flowers and fruit.

As with many kinds of animals, female plants often reach sexual maturity at a later age than males. This is not an important consideration in monoecious plants, those that produce both male and female reproductive organs on the same flower or plant, but it is in plants that are either male or female. These dioecious species include popular landscape plants such as hollies (*Ilex* spp.) and bayberries (*Morella* spp.). Male hollies will often begin producing flowers at least one year before similarly aged females, but that should not be surprising. It takes far more energy for a female holly to successfully flower and fruit than it takes for a male holly to simply produce pollen. To generalize, most herbaceous plants reach sexual maturity in one to two years if conditions are favorable; most woody shrubs and subcanopy trees do so in three to five years; and most canopy trees need at least seven years. If you are adding container-grown plants to your landscape, it may not be possible to know the exact age of the plant you are purchasing. The size of the plant can be used as a general guide, but the caliper of the main stem/trunk is more useful than the plant's overall height in estimating age. It is easy to generate a lot of upward stem growth by adding fertilizer so height can be a deceptive measure of age. Plants that are grafted may have flowering hastened or delayed depending on the type of rootstock on which the plant was grafted. In general, rootstocks that restrict growth (dwarf rootstock) produce plants that flower at a younger age in comparison to rootstocks that do not limit growth. This is commonly seen with dwarf fruit trees.

Age is only one of many factors that influence the production of flowers, cones, and sori (the spore-producing structures on ferns). Temperature, particularly cold temperature, plays an important part in the flowering of many plants. When winter temperatures drop extremely low, flower buds may be killed and the plant may flower sparsely. This condition is commonly associated with plants grown north of their normal geographical range or in places where winter minimum temperatures fall below those to which they are typically adapted. Such plants may bear flowers only on the parts protected by snow cover, such as thermal heat radiating from a wall or water. In below-freezing temperatures, the outer bracts of flower buds may be damaged and result in disfigured flowers that fail to open or set fruit.

On the other hand, a certain amount of cold temperature (usually at least as low as 45 degrees) is required for many plants to flower properly. Vernalization is the term applied to this cold-temperature requirement, and it is necessary for a great many ornamental plants commonly used in landscapes as well as many native plants. Plants are extremely good at detecting the right time of year to flower. The length of daylight is extremely significant, but temperature is also important. Plants will bloom later in the spring after a particularly protracted winter, and they will bloom earlier if spring temperatures arrive early.

Plants adapted to areas of the world that regularly have cold and warm seasons often can't sense the warm season accurately if they haven't first experienced the cold. For many spring-flowering trees and shrubs, at least six weeks of cold temperatures are necessary before the flower buds will break dormancy. Others will only bloom sparingly. Where I live, in west central Florida, many of my native and ornamental landscape plants flower far more profusely after a cold winter, even if we haven't had a hard freeze. Although many of my neighbors relish warm winters, I despise them. I can always wear a coat if it gets cold, but I have to wait another year to experience my flowers.

Some plants flower only in alternate years or even more sporadically than that. Most of these originate in habitats where acquiring enough nutrient energy to generate flowers is difficult. Where I live, oaks of the scrub and sandhill communities typically produce acorn crops only every three to five years while those common to moister habitats do so annually; this characteristic carries over even when both types are grown side by side in a landscape. It is genetically programmed.

Flowering also can be related to injury to the terminal leaders, either through browsing or by improper pruning. Many woody shrubs and or-

The onset of cold winter temperatures stimulates a great many plants to flower in the spring. Photograph by D. S. Damm.

namental trees produce flower buds in the fall that open the following spring. The loss of these buds after they are formed prevents flowering because the plant can't generate more until the next cycle. Examples of such shrubs are forsythia (*Forsythia* spp.), lilac (*Syringa* spp.), azalea, and some of the hydrangea (*Hydrangea* spp.) species. For this reason, it is recommended that pruning be done immediately after flowering. In most cases, this will ensure adequate time for new growth to occur and flower buds to be set on this new growth.

The nutritional balance of the plant also influences its production of flowers and other sexual structures. Nitrogen produces vigorous stem growth, but it comes at the expense of flower production (and the resulting fruit) if the balance is heavily skewed in favor of nitrogen over that of phosphorus and potassium. This is uncommon in nature or in natural landscapes, but it often occurs when inorganic fertilizers are applied excessively to stimulate growth. If the application of fertilizers is warranted, formulas more balanced between nitrogen and phosphorus/potassium should be used to balance growth and reproduction.

The production of flower buds occurs at different times of year in different species of plants. Azaleas (*Rhododendron* spp.), like this one, generally produce next spring's flower buds immediately after flowering.

Many environmental forces influence the onset of flowering, but none of this actually explains how plants interpret them and put this information to use. Plants respond to the outside world and make decisions about what to do, just as animals do. Over the past few decades, a large body of research has been developed to provide a very detailed understanding of how photoreception works in plants, but our understanding of how plants gather information on temperature is less clear. Obviously, temperature affects enzymatic reactions on a cellular level, but it would seem that plants have actual receptors that help them discern temperature fluctuations and use these receptors to respond. Recent research also suggests that light and temperature receptors work in unison, but how this actually works is the subject of much current plant research.

Directed responses to outside forces in both plants and animals are produced by hormones. That's the definition of a hormone: an organic chemical produced by a living organism that directs a response from an outside stimulus. Plant hormones will be discussed in much more detail in chapter 10, but one of the greatest botanical mysteries of the past century has centered on the hormonal control of flowering. Immature flowering plants do not produce flowers; they produce only new shoots, roots, and leaves. Research has isolated a hormone, produced in extremely small amounts, that actually triggers flowering, florigen. Florigen is a protein produced in the leaves of flowering plants that enters the phloem and gets transported to the meristematic tip of the stem. Here it acts as a sort of alarm clock, waking the cells to begin a complex set of reactions that culminate in the production of flower buds. Flowering is more complex, however, than florigen can accomplish on its own. It is not possible, for example, to make a plant flower simply by spraying it with florigen. Like all plant hormones that will be discussed in chapter 10, there is a suite of hormones that act together and opposite one another that serve as checks and balances to the overall production and development of flowers.

Much of the discussion above has been directed at flowering plants, but not all plants produce flowers for reproduction. It is a fairly recent phenomenon evolutionarily speaking, but it has been extremely successful, and our modern world (and our landscape palette) is comprised mostly of flowering plants. Biologists estimate that 70–80 percent of all living plant species produce flowers. Far fewer produce spores or seeds inside a cone than inside a fruit. Regardless, all plants use environmental cues to determine when to reproduce, must reach a critical age before they are capable of it, and regulate the production of their sexual structures by their internal hormonal system.

REPRODUCTIVE STRUCTURES—NONFLOWERING PLANTS

Mosses and their relatives (bryophytes) produce spores and gametes in specialized structures. As mature adults, they function either as males or females. The males produce sperm, and the females produce ova. In mosses, these structures occur at the tip of the stems; in liverworts and hornworts, they occur on the leafy thallus as outgrowths or inside slight concavities. Bryophyte sperm is flagellated and must swim across a film of water to reach the unfertilized egg in a neighboring female's ovary. Sometimes a raindrop will splash sperm from the male structure to a nearby female; sometimes the sperm merely swim to the ova when the conditions are right. Sexual reproduction in bryophytes can occur, therefore, only in

Mosses and their relatives reproduce with spores produced in specialized structures.

moist or rainy periods. It is a restrictive system, but it has worked well for millions of years.

Fertilization occurs when a sperm reaches an unfertilized egg. In mosses, fertilization results in the production of a stalked capsule at the top of the female parent, and thousands of spores are produced inside. Though liverworts and hornworts produce slightly differently shaped structures on the female parent, these too produce thousands of spores. Eventually these sexually produced structures mature and shed their spores, and the spores are carried on the wind, fall on the ground, and germinate when the conditions are moist and warm enough. Spores are lightweight and easy to produce. Mosses and their relatives occur nearly everywhere because they are so easily transported and because they are resilient, but their reproduction is completely dependent on moisture and outside weather forces.

Ferns and their relatives share the basic model of their bryophyte ancestors by maintaining spores as their primary mode of reproduction, with the
consequential reliance on water. They made one major change, however.
Adult ferns produce spores, whereas adult mosses produce gametes. Mature
ferns produce spores seasonally for the life of the plant. Unlike the tiny and
ephemeral spore-producing capsule of a bryophyte, mature ferns generally
live for years and are capable of producing millions of spores over their
lifetimes. The spores are produced in special structures called sporangia,
which are collected together in variously shaped units called sori, and these
are then normally covered by a wax-paper-like covering called an indusium.
This additional complexity gives added protection to the developing spores
that their bryophyte ancestors lack. All the advantages of producing spores
are still evident in ferns, as is the necessity for moist conditions to initiate
their germination.

Spores are not produced by sex but rather by the asexual stage. Sex ensures the mixing of gametes between two parents, which works to satisfy
the long-term need for genetic diversity. Pteridophytes (ferns and their
relatives), therefore, have to have a sexual stage that produces gametes.
When pteridophyte spores germinate, they develop into a tiny independent structure smaller than your little fingernail, the prothallus. One
end of the prothallus becomes anchored to the ground, a rock, or some
other solid structure, by small structures called rhizoids. Rhizoids are not
roots; they serve only to anchor this tiny nonvascular leafy prothallus to
the ground or other substrate. Near the rhizoids, male sperm-producing
structures develop, and at the other end, egg-producing structures are
formed. Each prothallus, then, produces both types of gametes, the sperm
are flagellated as they are in the bryophytes, and they also have to swim
through a thin film of water to reach an egg—either on the same prothallus or another one nearby. Once a sperm reaches an egg, the embryo
quickly develops a root and a true leaf (frond) and the prothallus beneath
it disintegrates.

Ferns and their relatives are extremely successful and occur nearly everywhere on earth, but they too are extremely dependent on moisture for
reproduction. Though their spores can remain viable for years after they
are produced and while waiting for the conditions needed for germination, once they germinate they have a short window to complete the rest of

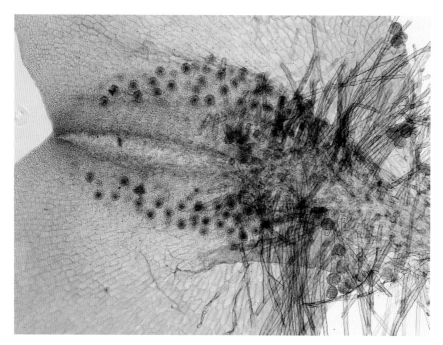

Ferns and their relatives produce eggs and sperm for sexual reproduction on tiny, short-lived plants called prothalli. Photograph by Mira Janjus.

their life cycle. Without ample moisture, the nonvascular prothallus quickly shrivels, and the gametes cannot reach one another.

Gymnosperms produce cones and seeds, and these strategies represent huge advances over those used by ferns and mosses. While spores can be produced in huge numbers with very minimal energy output by the parent plant, they provide virtually no nutrition to the embryo inside. Seeds are greater energy investments, but they nurture the embryo as it develops and germinates. While a high percentage of germinating spores do not survive into infancy, many germinating seeds do. The energy investment that the parents make in producing seeds pays off.

While ferns and bryophytes produce their spores in relatively fragile structures, gymnosperms surround their seeds with woody cones. Pollen is produced in male cones with small scale-like bracts, not very different in composition from the sori of ferns or the scales of the cone-like strobili of horsetails, spike mosses, and club mosses. Gymnosperm pollen is produced in huge numbers, and only a tiny fraction is needed to fertilize the ova in female cones. Pollen is expendable; eggs are not.

Gymnosperm eggs are produced in female cones that are far more substantive than the papery-thin scales of the male cones. The scales of female gymnosperm cones are thick and often spiky to help ward off herbivores. They open just wide enough to admit a few airborne pollen grains and then, once this occurs, close tightly. Enclosed in this dense structure, the pollen eventually finds the eggs and fertilizes them. This may take more than a year to occur. Once it has, the seeds with their tiny embryos begin to grow, and the cone around them expands. The maturation of these embryos also may take more than one year before they are ready for dispersal. Fully mature female cones then open their scales and release their seeds. By the time a mature pine, hemlock (*Tsuga canadensis*), fir (*Abies* spp.), cycad, or other gymnosperm sends its seeds out into the world, they are fully prepared to meet the obstacles in front of them.

Gymnosperms, like these pines (*Pinus* spp.), reproduce by cones. Small, short-lived male cones produce pollen, whereas the females produce ova that will eventually produce seeds inside a large "wooden" cone.

Seeds do not generally disperse as far from their parents as do spores, but such long-distance dispersal is not as significant a need. Few parents of any kind want their offspring to grow up in their shadow; the desire is to have them grow up independent and just far enough away that they don't forever tax the parents' resources. If a seed plant has matured and produced seeds, chances are that the conditions in the general vicinity of that plant will also meet the needs of its offspring. Over the years, populations will spread out until they meet conditions they can't overcome.

Most gymnosperms use wind to disperse their pollen and also to disperse their mature seeds. Lightweight pollen can disperse over great distances, just as spores do, but the mature seeds are relatively heavy. Most come equipped with a papery wing that aids in their dispersal; others, like those of yew (*Taxus* spp.), cedar, cycads, and gingko (*Gingko biloba*), become partially covered in a fleshy aril that often ripens to a bright color and attracts seed-eating wildlife. Birds generally do not kill the embryo inside the seed when they consume them. Once the aril is digested, the seed may be passed in the feces miles away from where it was eaten. Rodents can destroy the embryo if they consume the entire seed, but often they are more interested in eating the "fruity" aril off the seed. As many carry the seeds away from the parent plant before settling down to eat them, rodents and other mammals do their share of short-distance seed dispersal.

FLOWERING PLANTS

The gymnosperm approach to seeds worked really well up through the age of dinosaurs, but flowering plants quickly emerged following that period and there has been no turning back since. In today's world, there are more than 250,000 species of flowering plants, and they comprise about 80 percent of all the plant species on earth. The explosion of this approach to reproduction clearly demonstrates how effective it is. The process of evolution favors successful models.

Flowering plants continued with the same basic approach of producing seeds inside a protective structure, but they improved the model in several distinct ways; for many, that meant giving up on the vagaries of wind power to fertilize the embryo and disperse it once it matures. In most flowering plants, these important processes were usurped by animals, and this led to a

Grasses, like this lopsided Indiangrass (*Sorghastrum secundum*), produce flowers that are wind pollinated.

More than 80 percent of flowering plants are pollinated by animals such as bees and butterflies.

wild and crazy evolutionary race among pollinators and seed dispersers. In the Jurassic world populated by dinosaurs and made famous in film, there were plenty of insects, but no pollinating ones. There were no flowers to pollinate. There were a few true birds and mammals, just none that ate fruit. There was no fruit either.

In our modern world, there are still flowering plants that rely on the wind to disperse their pollen. The most obvious are the grasses, of which there are some eight thousand species worldwide, but only about 12 percent of all flowering plants rely on wind to disperse their pollen. The rest are pollinated by pollinators of some kind.

Most flowers are designed to be pollinated. Their structure is based solely on luring pollinators to them and maintaining that relationship by providing a reward for stopping by (or at least the pretense of a reward). Over the many years that flowering plants have been evolving, the relationship between flowers and pollinators has generally become more complex and targeted. Few flowers are open to all pollinating animals. Pollinators and their flowers are involved in a never-ending dance of continually increasing complexity.

Flower Structure

Flowers are composed of some combination of four main parts. Two of them are nonsexual, and two are there strictly for sex. The nonsexual parts are the leafy structures that enclose the bud (sepals) and the flower petals themselves. Flowers can lack three of the four parts and still be flowers. In fact, many flowers lack at least one of the four parts. They are called incomplete flowers. The only requirement is that they have at least one of the two sexual parts, the male stamens and/or the female pistils.

Sepals protect the developing flower when it's a bud. They grow as the other flower parts inside them develop, and they peel back when the other parts mature. Some sepals provide additional protection by having spines, thorns, or other defensive structures on their surface, but most are green and leafy, and they provide some photosynthesis benefits while they cover the more fragile flower parts inside the bud.

In most flowers, the sepals fold back and wither away when the petals and reproductive structures are mature. There is no need to retain them because their primary function is no longer important. Sometimes, however, the sepals change color and become petal-like. When this happens they are correctly called tepals. As tepals, they double the number of petal-like structures surrounding the sexual parts of the flower, and this increases the visual cues that attract potential pollinators. Wind-pollinated flowers don't waste time producing showy sepals or petals, but animal-pollinated flowers often do. Lilies (Liliaceae family), iris, and canna are all examples of flowers with sepals that become petal-like. All orchids have tepals. Since most flowers produce sepals anyway, keeping them around as false petals after the bud has opened costs the plant very little extra energy and can significantly increase the flower's ability to attract the attention of pollinators.

When we think of flowers, however, we are usually focused on the petals. They are the showy part, and they are the end result of much of our gardening efforts. Petals are all glam and minimal substance. They are a lure to attract the attention of pollinators. They are the red light hanging in a window that tells the rest of the world that the flower is open for business. In fact, many flowers don't even bother with them; they attract pollinators purely by scent or by modified leaves that substitute the petal's role.

Petals come in all styles and colors. The arrangement of the petals is tied closely to the type of pollinator it wishes to attract. Flowers with deep tubes

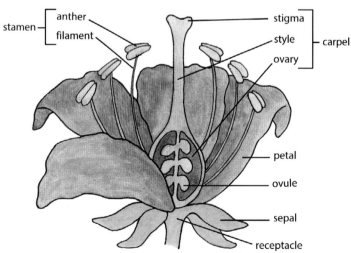

Structure of a flower.

require pollinators with long tongues or beaks to reach the nectar; petals that fold backward and out of the way of the reproductive parts can be accessed by those with shorter tongues and beaks. Truly complicated flowers, like those of some orchids, have evolved extremely close ties with a single species or group of closely related species to do the pollination. Flower shape is not random; it has evolved independently in all flowering plants to meet the needs of the pollinators to which it has chosen to link.

The color of the petals also is not random, though plant breeders routinely ignore this by creating colors that are not natural for the plant. Butterflies and birds are attracted to red and yellow colors, and these contrasting colors in the same flower are readily seen by a number of pollinators. It is no accident that hummingbirds are attracted to red- and orange-colored blooms or that many of these flowers are also used by butterflies. It's not that they won't pollinate flowers of other colors, but they prefer red and yellow and often visit these first.

Bees and pollinating wasps and flies have vision that is shifted toward the blue end of the visible spectrum. Flowers that have evolved to be mostly bee pollinated, therefore, have purple and bluish petals. Bees do not see red colors but do see into the near-ultraviolet. Thus, when a bee visits a brightly colored flower, even those that are red and yellow, they are cued most likely by a pattern they see in the ultraviolet portion of the spectrum. Many flowers have these nectar guides, and we see them as dots or dashes on the inner end of the petals, close to the reproductive parts. To our eyes, they are simply interesting markings, but under an ultraviolet light they assume a different role. Like the landing signal aboard an aircraft carrier, these markings direct bees to the nectar and facilitate pollination. They are not necessarily seeing the red or yellow petals; they are following the markings of the nectar guides.

Bright colors work exceptionally well during the day, but not at all at night. Our vision, and that of pollinators, shifts from a world directed by color during the day to one in black-and-white during the evening. Of course, black disappears in the darkness so flowers designed to be pollinated at night are white. Crystalline white flowers, without nectar guides at the base of the petals, are almost universally pollinated by moths or bats. Other types of pollinators are generally resting during these hours, so it is left to those that are active. Most moths and bats are not pollinators; only a specialized subset performs this type of service. Sphinx moths, for example, are the major group of moths that visit night-blooming white flowers. The tobacco hornworm that may plague your tomato plants is an invaluable pollinator as an adult moth. In nature, there are few things that don't play a positive and a negative role during their lifetimes, and sometimes it's only a matter of perception which is which.

The male sexual part of a flower is collectively called the stamen. It is a relatively simple structure composed of two parts: the filament, which is the stalk that holds the anther up above the petals and sepals; and the anther,

Flowering plants often go to great lengths to ensure that only a small subset of possible pollinators are involved. Night-scented and night-blooming flowers are white and attract the services of moths or bats.

which produces the pollen. The filaments vary in length and shape and may not be present at all. It's the anthers that are important as they produce pollen, and pollen contains the sperm that will ultimately fertilize an egg and produce an embryo.

Pollen is relatively cheap to produce compared to eggs, so each anther produces large numbers of pollen grains, and each flower contains anywhere from three to hundreds of anthers. The evolutionary trend has been toward reducing the number of anthers, not increasing them. As pollination has become more efficient, there is less need to produce as much pollen.

What we see as a simple grain of pollen is actually a bit more complex. Each grain is encased in a hard outer shell that protects the gametes inside until conditions are favorable for fertilization. The outer shell is often sculpted with an intricate pattern of spines, bumps, and indentations that is stunning to look at under a microscope and unique to each species. If pollen does not find a suitable environment in which to fertilize an egg, the hard outer shell can persist for thousands of years. Because each species has a unique shell pattern, it is then possible to examine ancient sediment deposits and make inferences as to what types of plants were common at different times in prehistory. It also allows forensic pathologists to help solve crimes if pollen from a crime scene comes from plants with very specific natural ranges or habitat conditions. The outer shell also contains grooves or pores (either one or three) that the sperm inside will use to leave the pollen grain on its journey to fertilize an egg.

Inside a pollen grain are two cells: the cell that produces the sperm; and the cell that develops a tube that burrows toward the ovary and the unfertilized eggs. The sperm needs this pollen tube to reach the egg and fertilize it.

Pollen that lands on the female part of the flower has a long path to navigate before fertilization can occur. Pollen initially lands on the tip of a stalk called the stigma, some distance above the ovules. The pollen tube cell starts to grow as soon as the pollen grain sticks to it, and the complex chemical reactions start to occur that tell it that it should start to grow. While all of this is happening, the sperm-producing cell inside begins to develop and the sperm mature. Two sperm cells are produced in each pollen grain. The tube cell burrows to the egg and the now-mature sperm arrive to fertilize it.

The female part of the flower is collectively called the pistil and is comprised of three separate parts: the stigma, style, and ovary. The stigma is the top of the pistil and the site where pollen lands and sticks. Stigmas come in

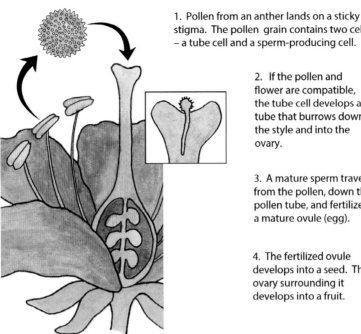

1. Pollen from an anther lands on a sticky stigma. The pollen grain contains two cells – a tube cell and a sperm-producing cell.

2. If the pollen and flower are compatible, the tube cell develops a tube that burrows down the style and into the ovary.

3. A mature sperm travels from the pollen, down the pollen tube, and fertilizes a mature ovule (egg).

4. The fertilized ovule develops into a seed. The ovary surrounding it develops into a fruit.

Pollination diagram.

a wide variety of shapes, but they function the same. When a flower is not quite mature enough to be fertilized, the stigma is slick and smooth so that pollen cannot stick to it. When the flower is ready, the stigma changes and becomes sticky so that any pollen that lands on it will be kept in place.

The style is the tube-like structure that connects the stigma to the egg-producing ovaries. The style is more than an impediment that the tube cell has to burrow through to accomplish fertilization of the eggs; it allows the female part of the flower to have some control over whom she mates with. Sex in plants is not as passive at it might seem to us. Plants do not necessarily mate with the sperm of the first pollen grain that sticks to the stigma, nor do they mate with the sperm of males at random.

The female portion of the flower has some ability to select her partner, and it is the style that stops incompatible pollen from penetrating it to reach ovary. When the pollen tube starts to extend within the style, genetic information is exchanged between the pollen and the plant. At the same time,

As with most of biology, exceptions to this simple structure abound. Some plants that are dioecious can switch sex depending on the conditions around them. Jack-in-the-pulpit (*Arisaema triphyllum*) and its relatives express sexual differences at different stages of growth; smaller plants produce all or mostly male flowers, but the male flowers are replaced by female flowers as the plant grows larger over the years. Some plant species have plants that produce more male flowers early in the year and more female flowers later in the season.

Not all monoecious plants are strictly one sex or the other. In reality, many produce a few flowers of the other sex. Such plants are called polygamodioecious, a winning word for a game of Scrabble® if there ever was one. It is not uncommon to find a few fruit on a male dioecious plant. The reason is that it is really polygamodioecious. Plants that produce male or female flowers on the same plant are still considered monoecious. Most conifers and birches (*Betula* spp.) do this.

Attracting Pollinators

More than 70 percent of all existing plants produce flowers that are pollinated by some combination of pollinating insects, birds, and mammals. Most draw the initial attention of their pollinators by a gaudy show of petals, but it takes more than that to bribe them into doing the deed; it takes giving them a substantial reward, or at least the image of one.

The overproduction of pollen is one method that flowers use to convince certain types of pollinators to visit. The pollen grain itself is a complex collection of fats, proteins, amino acids, vitamins, and carbohydrates in relatively high amounts. Humans have discovered that, and many of us supplement our own diets with honeybee pollen. Different plant species have different compositions of these compounds, and it is tied to the types of pollinators they are most associated with. Bees, of one kind or another, are responsible for 75–80 percent of all pollination services, and flowers reward them by providing pollen for their young. Pollen provides essential proteins and fats, and many could not survive or rear their young without this addition to their diet. Most bees and some pollinating bats have evolved complicated structures on their bodies to transport the pollen. Bumblebees and honeybees take away nearly 90 percent of all the pollen they collect, but the remaining grains get transferred to other flowers for pollination. Producing a lot of pollen is a necessity, not a wasteful exercise. Plants that

Pollen is an important food source for many bees. It is collected and then fed to their larvae.

are wind pollinated or not primarily pollinated by bees or bats (e.g., those pollinated by butterflies and moths) do not produce pollen that is as high in proteins, amino acids, vitamins, fats, and carbohydrates. In this case, it would be wasteful to do so, and evolution is anything but wasteful.

Not all pollinators are purposeful pollen collectors, but nearly all are drawn to floral nectar. Nectar is a sugar-rich liquid produced by plants in glands called nectaries, and floral nectaries are produced in small glands at the base of the filaments or sometimes at the base of the petals. Nectar collects in these small glands. Different flower shapes have different abilities to store nectar. Tubular flowers, for example, are better adapted to store floral nectar than other styles. That is why hummingbirds, with their high energy needs, are most drawn to tubular-shaped flowers.

Pollinating insects depend on nectar for its high energy content, but as a sugary compound it is quickly burned off, and this forces them to revisit throughout the day—which is exactly what the plant intends. This mutualistic relationship works for both sides. In some plants, the nectar also feeds predatory wasps that feed on caterpillars as well. In these cases, the production of nectar helps provide a defense that keeps the plant from being consumed.

Because evolution is economical, plants only produce nectar in the amounts and at the times of day best suited to ensure their maximum chance of pollination. Not all pollinators are active at the same time of day or during the same months of the year, so plants are not either when it comes to nectar production. The production of nectar is not free. The sugar content is more concentrated than that in the phloem so producing it takes resources away from growth and energy storage. Flowers stop the production of nectar once they are pollinated. It is an elegant system controlled by sensitive plant hormones.

Plants also regulate the production of nectar throughout the day depending on weather and visitation rates. It would be extremely wasteful, for example, to produce more than was necessary or to overfill a floral tube. Nectar removal stimulates nectar replacement. In one experiment using beardtongues (*Penstemon* spp.), researchers found that twice as much nectar was produced during the day if the nectar was removed hourly than if it was removed only at the beginning and end of a six-hour day. Additionally, hummingbird-pollinated species made more nectar than those primarily pollinated by bees regardless of number of visits during the day. Hummingbirds have greater appetites than bees; flowers more reliant on them have had to evolve larger single meals to keep their interest. This type of relationship between pollinators and nectar production has been shown in other flower species in other research studies. It simply makes evolutionary sense that floral nectaries have a mechanism to regulate the production of nectar by refilling their reservoirs only when nectar has been removed and to the extent that more is needed. Fluctuations in pollinator abundances and weather conditions would make an inflexible nectar secretion schedule less viable than a flexible one.

Most flowers also produce olfactory clues to attract attention. This is more important for insect- and mammal-pollinated plants than it is for those that rely on birds. It also is the only effective method for plants with small or difficult-to-see flowers. Birds, with very few exceptions, have virtually no sense of smell, and hummingbird-pollinated flowers are rarely fragrant. Birds use vision for long-distance cues; insects have very poor distance vision but very keen abilities to detect fragrance. Bat pollinators also have to rely more on their keen sense of smell as they move about in darkness.

Scent is typically a complex mixture of low molecular weight compounds emitted into the atmosphere, and the chemical structure of these compounds is closely tuned to attract the specific pollinators the plant most desires. Although the flowers of different species can be nearly identical in their color or shape, there are no two floral scents that are exactly the same because of the large diversity of volatile compounds and their relative abundances and interactions. Thus, scent is a signal that directs specific pollinators to a particular flower whose nectar and/or pollen is the reward. Species pollinated by bees and flies have sweet scents, whereas those pollinated by beetles have strong musty, spicy, or fruity odors. The fragrance of a flower is a finely tuned evolutionary response, not merely a horticultural curiosity, though our modern plant breeding programs have often disrupted these relationships by breeding for floral structure instead of fragrance. The

Flowers produce nectar as a lure to feed adult pollinators. By doing so, their pollen is picked up and carried to other flowers.

garden roses (*Rosa* spp.) of my childhood that once smelled so heavenly have largely been replaced by blooms that only look pretty. This is partially the result of plant breeding programs designed to maximize the "life" of cut flowers. The production of fragrance has a metabolic cost to the flower. While fragrance increases the flower's chance of being pollinated, it makes the flower wilt sooner. Everything being equal, a fragrant rose will wilt in a vase well before a nonfragrant one.

Plants can regulate the production of fragrance and tend to maximize it only when the flowers are ready for pollination and when its potential pollinators are active. The physical condition of the plant also affects the quality and production of fragrance. Plants that maximize their output during the day are primarily pollinated by bees or butterflies, whereas those that release their fragrance mostly at night are pollinated by moths and bats. During flower development, newly opened and immature flowers—those not yet ready to function as pollen donors—produce fewer odors and are less attractive to pollinators than are mature flowers. Once a flower has been sufficiently pollinated, quantitative and/or qualitative changes to the floral bouquets lead to a lower attractiveness of these flowers and help to direct pollinators to unpollinated flowers instead, thereby maximizing the reproductive success of the plant.

The production of fragrance is complex and comes largely from the metabolism of various cell membranes in the flower triggered by enzymes that serve to break the membranes down. Fragrance can come from different parts of the flower. Some flowers have fragrant pollen, but it also can originate from parts of the petals and pistils. Often it is produced in multiple locations, and the various fragrances mix with one another. Rose pollen, for example, contains a fragrance profile separate from the rest of the flower, but fragrances from the rest of the flower usually overpower pollen fragrances.

The volatile essential oils released by this process evaporate and combine with one another to produce the distinctive fragrances recognized by the pollinators and ourselves. Different species of plants are more complex than others, depending on how complex their relationship is to specific pollinating insects. Orchids, for example may produce one hundred different volatile compounds, whereas the garden snapdragon (*Antirrhinum majus*) produces seven to ten.

Fragrance can be a lure that provides a tangible benefit to its pollinators. Sometimes flowers produce compounds that are narcotics, and insects are lured and drugged until pollination is complete. Various orchids, among some others, produce sex pheromones that attract male insects looking for a mate or compounds they need to acquire to attract a female. Other flowers produce products such as waxes and pheromones that insects store in their bodies to make them repellent to other predators. Though we use fragrance as an added benefit to the landscape, plants have a far more complex relationship with it.

Fruit

Only flowering plants make fruit. It is an evolutionary advancement that we often take for granted because it *is* so successful, but producing fruit is a modern-day miracle that now provides a great amount of the food that other life on earth consumes. The spores of mosses and ferns have little nutritional value, and the seeds of gymnosperms cannot compare with the overall value that the fruit and seeds of flowering plants provide. For many modern-day animals, foliage is difficult or nearly impossible to fully digest. Fruit, and the seeds inside, are consumed by nearly everything, even by most of the world's carnivores.

Fruit are the result of pollination in flowering plants. The sperm from the pollen grain fertilizes the ovule located in the flower's ovary. When this happens, it triggers a reaction that causes the ovary to expand and grow around the fertilized seeds. The ovary develops into a fruit, the fruit provides protection for the developing embryos, and its maturation coincides with the maturation of the embryos so that they are released and dispersed at the proper time. It is the most elegant system in the world of plants.

Fruit development involves a complex interplay of cell division, differentiation, and expansion of plant tissues and is carefully coordinated temporally and spatially. Plant hormones regulate these processes throughout and lead to mature fruit and viable mature seed. More than a half-dozen plant hormones are known to be involved at various stages of fruit development, but the entire process is exceedingly complex, and researchers are still trying to untangle all the various steps and chemical processes involved.

Fruit development can generally be considered to occur in four phases: (1) fruit set; (2) a period of rapid cell division; (3) a cell expansion phase; and

The evolution of fruit was a major step forward in the overall evolution of modern plants.

(4) ripening/maturation. Fruit set involves the decision whether to abort the ovary or proceed with fruit development. As gardeners, we know that many flowers fail to produce fruit and that some fruit are aborted by the plant shortly after they are formed. This is especially true for young plants recently added to the landscape or for plants that seem to be in decline.

Fruit set is normally dependent on pollination. Pollen triggers fruit development, indicating that positive signals are generated during pollination. In the absence of the signals generated by pollination, the flowers fall off, and no evidence of fruit formation occurs. Plant biologists have discovered that certain hormones are involved at this stage and that fruit can be induced to form from a flower when these hormones are applied to the stigma, even when no pollen has been applied. However, most plants can produce fruit even when they have a mutation that makes them deficient in these hormones. The mechanism for fruit production is more complex than the simple presence of one or two major plant hormones.

Temperature also affects fruit set after the flower is pollinated. Pollination is the arrival of pollen grains to the tip of the stigma. This triggers the initial reactions involved with the development of the fruit. Fertilization, however, occurs when the sperm has traveled down the pollen tube and combined with the ovule. This phase is partially controlled by temperature. Cool temperatures at the time of pollination slow the growth of the pollen tube and thus the arrival of the sperm. Ovules and sperm do not remain viable for long, and either one or both can perish before fertilization occurs if the process takes too long. For most plants, cool temperatures reduce fruit set; warmer temperatures increase it. There is a limit, however, to how high the temperatures can be to sustain fertilization. Plants are not adapted to pollination temperatures above a certain maximum, and fruit set declines rapidly if these temperatures are reached. For many nontropical plants, the most successful fruit set occurs at temperatures between 70–75° Fahrenheit (about 20 to 25° C). At higher temperatures, individual floral buds and open flowers start to be aborted.

Last, the condition of the plant can influence fruit set. Plants in failing health or those deficient in essential nutrients may flower but not set fruit. The process of producing fruit is very taxing on plants. If they cannot replace those nutrients lost from last year's fruit production, they often will fail to set fruit in subsequent years until those deficiencies are made up. Nitrogen, for example, has been used to increase fruit set in various plants. As one of the most significant elements responsible for plant growth, maintaining a nitrogen balance is critical. Research on olives, for example, has shown that the addition of nitrogen not only increases fruit yield but also influences the production of female flowers over that of male flowers.

The addition of phosphorus has rarely been used to enhance fruit set in agricultural crops, but potassium is critical. Fruit typically has a high concentration of potassium; in some plants more than 60 percent of their entire biomass is concentrated in the fruit. Potassium deficiencies will severely diminish fruit production, and increased potassium availability will enhance it.

Of all the micronutrients plants require, boron seems to be the major element also required for normal fruit set. Plants deficient in boron often

Not all pollinated flowers eventually produce fruit. Many are aborted before they fully mature.

produce imperfect flowers; flowers that typically have both male and female parts are replaced by flowers that lack the female component. Research also has shown that boron-deficient plants fail to set fruit properly. It has been known for a long time that boron is essential for the successful growth of the pollen tube through the stigma, style, and ovary to the ovule; and for the mitotic divisions necessary to produce the sperm. Though boron is not commonly deficient in most soils, its deficiency should be suspected if other factors described above are not likely causes of a persistent problem with fruit set.

Fruit also can be dropped partway through the development process. This occurs mostly because the plant doesn't have enough energy to ripen all the fruit it is currently carrying. In a sense, the ripening fruit are competing with one another for the limited resources available to its parent. Aborting some of its fruit is a way to ensure that the others will receive the nutrition they need to fully develop. It's better to produce fewer fruit that contain fully developed seeds than it is to produce more with seeds unlikely to successfully germinate.

Once the fruit has set, continued development usually relies on the continued presence of developing seeds. The seed itself is responsible for the continued development of the fruit. If the seed is aborted naturally by the plant or if it is physically removed, fruit development is abruptly stopped. This is because the developing seed is producing hormones that are required for the fruit to continue its development. For example, the removal of strawberry seeds (actually what we think of as seeds in a strawberry are its real fruit; the red "fruit" is really something else) prevents the development of the red fleshy "fruit," but if the plant hormone auxin is applied following seed removal, fruit development continues. Commercial crops that produce seedless fruits, such as bananas, often show quantitative or qualitative differences in hormones in the ovary compared to seeded varieties. Seedless varieties are not a natural strategy; they can persist only with the intervention of modern horticulture, and they arise because of an imbalance in the production of hormones by the plant that circumvents what should be the role of the seeds themselves.

The phase of rapid cell division in the fruit involves all of its growing parts but is largely controlled by the developing seeds. The number of fertilized ovules in a fruit is correlated with both the initial cell division rate and the final size of the fruit. Fruits with an uneven distribution of seeds are often lopsided. There is a correlation between the production of the plant hormones by the developing embryos and cell division in the surrounding fruit tissues, but there is no direct evidence that the embryo directly regulates fruit cell division. As with so many things, the exact role of plant hormones in the growth and development of fruit is not known.

The cell division phase gradually shifts into the cell expansion phase. The rate and duration of cell division varies among fruits and also among tissues within a fruit. Tissues made up of many small cells at maturity continue dividing, whereas tissues composed of large cells stop dividing and start to expand. In tomatoes, the cell division phase lasts approximately seven to ten days, whereas cell expansion lasts six to seven weeks. Cell expansion accounts for the largest increase in fruit volume, often a hundred-fold increase or more. The plant hormones associated with fruit expansion are produced by the developing seeds. Plant biologists believe that a different hormone, also produced by the seeds, regulates cell expansion of the fruit, but this has not been completely documented. Application of this hormone does not always compensate for seed removal, and fruit growth is normal when this hormone is deficient.

Ripening represents the shift from the protective function to dispersal function of the fruit. Ripening occurs synchronously with seed and embryo maturation. In dry fruits, such as nuts and grains, ripening consists of desiccation, and fully ripe fruit are dry and hard. Ripening in fleshy fruits, like tomatoes and apples, is designed to make the fruit appealing to animals that eat the fruit and will thereby disperse the seed. Ripening involves the softening, increased juiciness and sweetness, and color changes of the fruit. All of these changes are a calculated effort to make them appealing to fruit-eating animals. It takes a great amount of energy to make these changes, but the rewards are worth it, as evidenced by how many fruit have evolved this strategy.

Ripening has been most intensively studied in tomatoes. A plant hormone, ethylene, that is not involved with growth and expansion is the major regulator of the ripening process. Inhibiting the natural production of it blocks ripening, and exposing unripened fruit to it causes them to ripen even though they are not fully mature. The process, however, is not quite that simple because fruit does not attain the ability to respond to ethylene until near the end of the cell expansion phase (the mature green stage). This part is regulated by certain genes carried in normal fruit-producing plants. Ethylene production is stimulated by exposure to ethylene; the initial production of ethylene by the ripening fruit stimulates the synthesis of more ethylene. It is a snowball effect until the fruit is fully ripe.

Fruit softening involves a partial breakdown of cell walls. Several enzymes are known to be involved in this process. The gene for this enzyme

Fruit, like these wild coffee berries (*Psychotria nervosa*), often change color when fully ripe. Unripe fruit are camouflaged against premature feeding by animals, whereas ripe fruit advertises to them.

is ethylene inducible; as the fruit produces ethylene, its production stimulates the production of the enzymes responsible for fruit ripening. At the same time, ethylene causes the production of different enzymes that cause a change in the production of various pigments in the skin. The carotenoid pigments (the same ones involved in photosynthesis) become noticeable, and the ripe fruit assumes a color from yellow to orange and red. Blues and purples come from a different type of pigment, flavonoids, which are not involved in other plant functions.

While all fruit develops from the flower's ovary, not all ovaries are similar and not all flowers produce ovaries the same way. A single pistil is also known as a carpal, but a flower can have multiple carpals fused into a single pistil. In this case, the ovary has chambers identical in number to the number of fused carpals. Fruits with a single seed or pit—cherries and plums (*Prunus* spp.) and avocadoes (*Persea americana*)—have one carpal. Fruits like cucumbers have three obvious fused carpals. Some fruits have more than three carpals. Good examples of these are members of the citrus family; each section of a mature orange (*Citrus × sinensis*) or grapefruit (*Citrus × paradisi*) actually began its existence as a flower carpal.

The number of seeds in a carpal is merely a reflection of the number of ovules originally present in each ovary. A grape, for example, is a many-seeded fruit that was a single carpal. A tomato, however, has many seeds in each of its multiple carpals. Cucumbers and other squash have three distinct carpal "lines" of many seeds per line. Each of these lines was a carpal in the original flower.

The outside of the fruit is called the pericarp and develops from the ovary wall. The pericarp can be dry and papery, like in maple (*Acer* spp.) or dandelion, woody like in acorns and hickory nuts, or fleshy as in berries (grapes and tomatoes) and stone fruits (cherries and peaches). Fruit types can be divided into two major categories: fruit that is fleshy when ripe; and fruit that is dry when ripe. Dry fruit can be further divided into those that split open when ripe; and those that don't. Beans and peas, for example, split when ripe and scatter their seeds. Acorns and other nuts do not. Pericarp differences reflect adaptations to different dispersal mechanisms (e.g., wind for papery pericarps, animal consumption for fleshy fruits).

Beans and peas are fruit, but the seeds inside are not; they are seeds. True nuts are fruit, but many of the things we call nuts are really just the seeds within a non-nut fruit. In fact, in a container labeled "mixed nuts," most of the contents are not nuts. Almonds (*Prunus dulcis*) and pistachios (*Pistacia vera*) are actually the single seed inside the pit of the real fruit of these trees, just as if we cracked open a peach pit and extracted the seed inside. Cashews (*Anacardium occidentale*) and Brazil nuts (*Bertholletia excelsa*) also are not nuts but rather the seeds produced inside the fruit of these flowering trees. Peanuts (*Arachis hypogaea*), of course, are a legume just like a bean or pea. The unshelled peanut is the fruit; what we consume is the seed inside.

Similarly, many fruit we consider berries are not technically so. Berries are soft-skinned fruit that often contain many seeds in one-to-many carpals. Tomatoes and peppers are berries. So are bananas, grapes, and kiwi fruit

A fruit is a ripened ovary. As such, fruit comes in a great many forms and structures.

(*Actinidia deliciosa*). Raspberries and blackberries are not really berries but rather a collection of many single-seeded fruit. The flowers of these types of plants have a large number of unfused carpals with a single ovule in each. Each fertilized ovule forms its own fruit (similar to the fruit of a cherry or plum), and these are all collected at the base of the single flower. Because of this, it is not uncommon to find raspberry and blackberry fruit with not all of the individual fruit fully formed. Such fruit are technically called aggregate fruit as the individual fruit of one flower are aggregated together into one structure.

Other fruit form additional fleshy material around the pericarp to further encourage its attractiveness to fruit dispersers. This additional material is termed "accessory fruit." Strawberries and apples are examples of accessory fruits, where some of the fleshy tissue is derived from flower parts other than the ovary. The true fruits of a strawberry are the tiny seeds outside the red fleshy part. The fleshy part that we eat develops from the original base of the flower, the receptacle. Most of the fleshy tissue in apples and pears (*Pyrus* spp.) develops from the hypanthium, which is a region of the flower

The evolution of seeds greatly enhanced the ability of plants to direct the dispersal of their offspring.

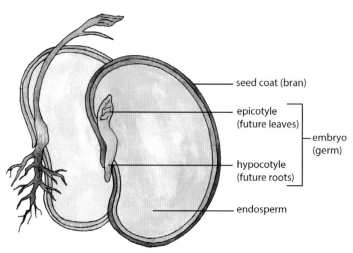

Seed cross section.

seed coat (bran)

epicotyle
(future leaves)

hypocotyle
(future roots)

endosperm

embryo
(germ)

SEED COAT

The seed coat is the seed's primary defense against adverse environmental conditions. A hard seed coat protects the seed not only from mechanical stress but also from microorganism invasion and from temperature and humidity fluctuations during storage. Chemical compounds in the seed coat contribute to seed hardness and inhibition of microorganism growth. During germination, the seed coat protects the seed from hydration stress and electrolyte leakage.

Hard seed coats that are impermeable to air and water induce a type of dormancy, called "seed-coat dormancy," by restricting the embryo's access to air and water. Embryos can't grow without air and water, so the seed remains dormant until the seed coat is penetrated. This protects the embryo by preventing germination before the time is right.

In nature, most seed coats require some kind of force to break them down, and the nature of this force is closely tied to the ecology of the plant and the conditions to which it is adapted. This process is referred to as stratification. Without stratification, seeds would germinate at times when the embryo inside could not grow and survive. It would do no good, for example, if red maple seeds in Wisconsin sprouted in January beneath two feet of snow in temperatures below freezing.

Stratification comes in many forms. Many plants adapted to cold winters require cold temperatures for their seeds to germinate. Even in west central Florida, where I live, there are native plants whose seeds require cold stratification. To ensure their germination, I put them on a moist paper towel in a sealed plastic container and store it in my refrigerator for at least thirty days. If I don't expose them to cold, they will not sprout.

Most often, the seeds of cold-stratified plants are dispersed during the fall. The arrival of cold winter temperatures is a trigger that tells them that everything is as it should be. Without it, the seed remains in a state of confusion, waiting for winter to arrive. Cold-stratified seeds have evolved not to "trust" the warmer spring temperatures unless they've first been exposed to the winter cold. This has great adaptive value for plants living in these types of climates as germinating too soon would guarantee the death of the newly emerged plant. Extended cold of sufficient length, followed by a sufficient period of warmer weather, helps break down the seed coat and triggers germination.

Plants with tiny seeds are often stratified by light. If you've ever sown lettuce (*Lactuca sativa*) or carrot seed, you quickly learn that seeds sown too deep beneath the soil surface do not germinate. Light stratification ensures that the germinating seed will not exhaust its energy supply before the newly emerging seedling can be self-sufficient. Small seeds have far less energy stored inside them than larger seeds. It's that energy that fuels the

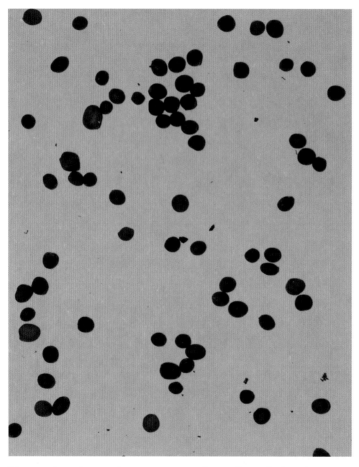

Most seed-producing plants require specific conditions before their seeds will germinate. Small seeds, like these mustards (*Brassica* spp.), cannot sprout if buried too deep. They require sunlight to germinate.

development of the seedling as it struggles to produce leaves and stems for photosynthesis and roots for water and nutrient intake.

Many weedy plants also require light stratification. Weeds are not the "right plant in the wrong place," but a class of plants characterized by an ability to rapidly colonize disturbed sites. Seeds of many weeds remain viable in the soil for long periods, waiting for soil disturbance. Remove the surface layer of turf that may be present on your property and scuff it up a bit with a tiller and see what emerges almost immediately. The first to emerge will, in all likelihood, be species requiring light stratification, and the vast majority will likely be weeds.

Some seeds require extreme heat to break their dormancy. This is especially true in areas where fire is a regular natural phenomenon. Fire resets the plant-succession clock, removes shade from the ground, and provides elbow room for more diminutive species to thrive in areas where larger/woodier species may have been dominating them. Seeds that require high heat to germinate are largely evolved to reside in habitats that regularly burn. By germinating only after a fire, the seedlings emerge with a maximum amount of available sunlight, a bit of potash fertilizer, and sufficient time to become established before the canopy is reestablished.

Other seeds are equipped with a seed coat that has chemical inhibitors that have to be removed before germination can be initiated. The outer oily seed covering (an aril) found on various cycads greatly inhibits germination, for example. When rodents feed on that aril and remove it, the seeds germinate in significantly less time than if the seed is left untouched and the aril slowly rots away. Other chemical inhibitors found on the seed coat are removed after they are washed away. This strategy ensures that sufficient moisture is available to the germinating embryo and is a common strategy of desert plants where rainfall is rare and unpredictable.

The other major stratification method involves physical and/or chemical abrasion. Such seeds have generally evolved to germinate only after passing through the gut of an animal. Digestive chemicals break down the seed coat of some of these plants. In others, the grinding action of a bird's gizzard or the nicking of the seed coat by the teeth of a rodent is what is required. Animals can carry seeds great distances before the seeds leave their digestive tracts. This ensures that the new seedlings will not be competing with their parents for resources, and it means that they will be deposited amid

Many fruit, like these flatwoods plums (*Prunus umbellata*), fall beneath the canopy of their parents, where they would eventually compete for resources. Such fruit are designed to be transported away by animal vectors.

a small mass of organic fertilizer. Seeds that are not consumed by wildlife fall directly beneath their parents. By not germinating without first being eaten, both the parents and their young are protected from the pressure of competition.

These types of plants often evolve complex relationships with the animals that disperse their seeds. One of the best examples involves the relationship formed between the dodo, a large flightless pigeon, and the *Calvaria major* tree on the island of Mauritius. The dodo was hunted to extinction by European sailors in the late 1600s. One of the dodo's major foods was the fruit of *Calvaria major*, a large, long-lived tree endemic to this island. After the dodo's demise, there were no other birds large enough to consume this fruit, and for the next three hundred years no juvenile trees were found on the island though the trees flowered and produced fruit annually. Dr. Stanley Temple unraveled this situation by feeding some of the fruit to large non-native pigeons. After passing through the gizzards of these birds, the seeds germinated for the first time in more than three centuries.

Nature has a plan for breaking through seed coats, but gardeners can speed the process by knowing which seeds require what types of stratification. Manually abrading or softening the seed coat is called scarification. You can scarify seeds by nicking them with nail clippers or a sharp knife, or abrading them with sandpaper. Take care that you don't damage the internal embryo when nicking or abrading seed coats. Other seeds will respond if they are soaked for a period prior to planting, chilled in the refrigerator, or put in near-boiling water.

Once stratification has occurred, water can enter the seed through the seed coat. This process is called imbibition, and it causes the seed to swell. Imbibition precedes germination and is required. Prior to this, the seed coat has prevented air and water from entering the seed, and the embryo has been in a state of suspended animation. Active metabolism inside the seed by the embryo resumes with imbibition.

Once imbibition occurs, the now-active embryo begins producing hormones that stimulate the synthesis of a suite of enzymes that start to break down the starches, proteins, and fats in the seed's endosperm and make them available to the growing seedling. This conversion of stored food into more accessible high-energy foods such as glucose and sucrose is critical to the germinating embryo.

ENDOSPERM

Fertilization of the ovule also produces endosperm, though it is produced a bit differently in gymnosperms than in flowering plants. Endosperm is the fuel source for the developing embryo. It also is the part of the seed that makes it so nutritious for seed-eating animals, including ourselves. Most vegetarians could not achieve a balanced diet without consuming seeds such as soybeans (*Glycine max*) and other legumes as well as cereal grains such as wheat (*Triticum aestivum*), rye, corn (*Zea mays*), and rice (*Oryza sativa*). The relatively high levels of proteins, oils, and carbohydrates inside these seeds are concentrated in the endosperm.

In cereal grains, it is the endosperm that produces the flour we consume. The seed coat is often polished off or is retained as "bran," while the embryo is referred to as the "germ." That is why all three components of a seed have such different nutritional values and why gluten-intolerant people can eat

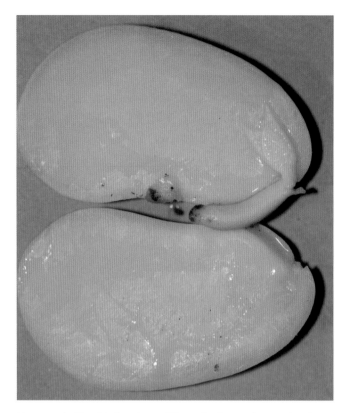

Most seeds, like this lima bean (*Phaseolus lunatus*), contain a large amount of nutritious endosperm to feed the developing embryo.

wheat bran and germ. The glutens are concentrated in the endosperm and used to feed the developing embryo.

Of course, glutens are a specialized protein found in wheat, rye, and barley (*Hordeum vulgare*). The seeds of different plants contain different types and amounts of proteins, fats, and carbohydrates. The nutritional value of a seed's endosperm varies greatly between species. Popcorn (*Zea mays* var. *everta*) "pops" because it produces two layers of endosperm, a moist core surrounded by a dry outer one. When exposed to high heat, the moisture inside causes the entire seed to burst and fold outward. Of course, popcorn and other types of corn are low in calories and fats and in this are much different from peanuts and soy. Coconut (*Cocos nucifera*) is actually the endosperm of the seed of the coconut palm. What makes Brazil nuts, almonds, pistachios, and cashews so nutritious is that we are eating the seeds of these plants (not the entire fruit), and their endosperm is especially high in oils, proteins, and carbohydrates.

Before imbibition, the proteins, fats, and carbohydrates are in storage. The dormant embryo does not need to grow. With imbibition, the endosperm starts to function as the embryo's energy supply. What makes seeds in our diet significant to us and other seed-eating animals is absolutely critical to the birth of the developing embryo. It is no lucky accident that most seeds are so highly nutritious.

The structure of the endosperm in the mature seed varies considerably between different species. The tomato contains a hard and thick endosperm cell layer in the mature seed, which undergoes extensive weakening during germination. In these seeds, the endosperm is largely used by the germinated embryo as it develops further. In contrast, seeds such as soybean and pea contain very little or no endosperm in the mature seed, as it is fully consumed by the embryo during seed development and prior to germination.

The endosperm also protects the embryo and controls embryo growth by acting as a mechanical barrier during seed development and germination. The embryo of a dry seed is embedded in the endosperm, and this shields it from damage from climactic events as well as the damage caused by chewing and other abrasion. Seeds can still germinate if some of the endosperm has been chewed away as long as the embryo has not been damaged. This would reduce the amount of nutrients available to it after imbibition and put the seedling at risk of not fully developing, but the embryo often can still survive.

Recent research has provided new insights into the regulatory functions of the endosperm during seed germination. Recent advances in seed biology have shown that the endosperm itself is capable of sensing environmental signals, such as light and temperature, and can produce and secrete signals to regulate the growth of the embryo. This indicates that the endosperm is not merely a nutrient source or a mechanical barrier to germination, controlled by the embryo, but that it also directs embryonic growth by actively secreting signals of its own. Germination involves complex back-and-forth communication between the embryo and its endosperm.

EMBRYOS

Plant embryos are baby plants. They may be structured a bit differently in different kinds of plants, but they all are baby plants formed by the union of sperm and egg. As in animals, plant embryos are sentient beings. They are aware of their surroundings, they direct their own development and eventual birth, and they grow and change prior to germination. Vegetarians sometimes lose sight of these facts in assigning different values between animal and plant life. And, though I do not presume to argue the morality of taking a plant life over an animal one, it cannot be argued that seeds do not contain baby plants and that baby plants are not alive and aware of their status and surroundings.

From the time of conception, plant embryos inside seeds contain primordia for their leaves and roots as well as an embryonic stem that connects

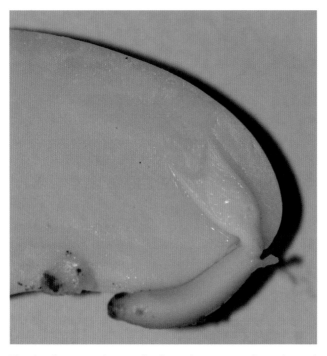

The developing embryo is clearly evident even in the early seed stage of this lima bean (*Phaseolus lunatus*).

them. Flowering plants contain one to two cotyledons. These are the organs that serve as an intermediary between the nutrition in the endosperm and the developing embryo, as well as often serving as the embryo's first leaf/leaves at germination.

The cotyledon comprises a significant part of the embryo prior to germination. The number of cotyledons in flowering plants is a major characteristic used to classify flowering plants into their major groups. Plants with a single cotyledon are referred to as monocots. Only about 25 percent of flowering plants are monocots, but they contain some very significant families. Grasses (including our major cereal grains), orchids, and palms are monocots. So are gingers (*Zingiber* spp.), bananas, irises, and lilies (*Lilium* spp.). Dicots comprise the remainder of the flowering plants. As their name implies, their embryos have two cotyledons.

Cotyledons may be epigeal (expanding on the germination of the seed, throwing off the seed shell, rising above the ground, and becoming photosynthetic) or hypogeal (not expanding, remaining below ground, and not becoming photosynthetic). Hypogeal cotyledons are common in plants, such as nuts, that rely heavily on their cotyledons as storage organs while the young plant is growing. Hypogeal plants normally have significantly larger seeds than epigeal ones. They are also capable of surviving if the seedling is clipped off because the seed and its developing buds are underground and undamaged. In epigeal plants, like beans, clipping the above-ground portion removes the meristem, and the newly developing root is incapable of producing a new stem and leaves. Epigeal plants are killed if grazed at this stage. Large, hypogeal seeds are energetically costly to produce. The trade-off is balanced by the relative risk the seedling faces during its first few days of being clipped or damaged.

The function of the cotyledon(s) varies somewhat between monocots and dicots. The cotyledon of most monocots is a highly modified leaf composed of a scutellum and a coleoptile. The scutellum is a tissue within the seed that is specialized to absorb stored food from the adjacent endosperm, whereas the coleoptile is a protective cap that covers the precursor to the stem and leaves. If you've ever sprouted corn or wheat, you've noticed that the initial spear-shaped leaf emerges from the grain without any extraneous extra tissues. The developing embryo continues to feed off the endosperm stored in the seed, but the endosperm remains inside that seed and is not noticeable.

Epigeal seeds, like this milkweed (*Asclepias* spp.), put all their energy, once sprouted, into the juvenile plant.

This differs in dicots. In most, the cotyledons emerge with the sprouting plant and are attached to the main stem. As the plant develops, this pair of cotyledons transfers the energy it has taken from the endosperm to the young plant as the plant works to develop functioning leaves and roots. As this happens, the cotyledons wither and eventually drop off, their mission accomplished. In many dicots, the cotyledons are also initially green and photosynthetic. During the first few days after germination, the cotyledons function as leaves until the true leaves develop from the tip of the germinated stem.

Gymnosperm seedlings also have cotyledons, but they vary in number from two to twenty-four and often form a whirl at the top of the emerging stem. The number of cotyledons can even vary within a single species,

Once the leaves emerge in this lima bean (*Phaseolus lunatus*) seedling, the energy left in the endosperm of the seed helps fuel its early growth before the roots and leaves are fully functional.

unlike the fixed number present in flowering plants. Like many flowering plants, gymnosperm cotyledons emerge above ground with the seedling and function as its first leaves.

The embryonic shoot occurs at the tip of the embryo and is referred to as the "shoot apical meristem" (SAM). Inside the seed, the SAM looks like a tiny bump of little importance, but it fuels all the upward growth that will occur with germination. The embryonic stem first elongates inside the seed prior to germination. As it develops, it also forms the plant's first leaves.

At the opposite end of the embryo is the other region of cell growth, the root apical meristem (RAM). The embryonic root begins to develop inside the seed and quickly emerges and matures following germination.

The development of the SAM and RAM inside the embryo is complex and is directed by a suite of hormones under the guidance of the developing embryo. Plant embryos begin development upon fertilization, and the initial structures are evident very early. This growth ceases for a time once these initial structures are formed, and the embryo enters a resting stage until imbibition. There is limited space inside the seed and no need to continue growth until germination seems imminent. When the conditions are finally met to favor successful germination, the seed swells and the embryo grows rapidly to keep pace with the expanding available space.

10

Plant Hormones

Hormones are organic chemicals, produced by plants or animals, that act as signals to elicit responses. In plants, as in animals, hormones regulate growth and development, but they also can change behaviors and cause movements. As humans, we observe the various ways our hormones affect our day-to-day lives. Our hormones affect our interest in sex, our fear of unexpected noises in the dark, and our ability to feel peace on a perfect spring afternoon. Plants are no different. They produce, and are influenced by, a wide assortment of hormones, and everything in their daily routine is directly controlled by them. Understanding something of these hormones is absolutely important if we are to understand how plants work.

Plant growth is ultimately determined by genetics, but it is influenced by the environment in which the plant lives and the hormones it produces. The expression of a plant's full potential for growth and development is under the control of its hormones. Most dwarf plants, for example, are individuals that either lack the normal contingent of growth hormones found in the rest of the population or they are individuals that lack the receptor that recognizes this hormone. In modern horticulture, we often select such aberrant individuals for landscape situations where full-sized specimens would be too large, but they can only be maintained by cuttings and tissue culture as the gene is recessive. In my part of the world, dwarf yaupon holly (*Ilex vomitoria* 'Nana') is invariably male. Though it does not produce fruit for the birds, I welcome it in my landscape because it is the perfect pollinator for my full-sized yaupon holly females. As a dwarf, it doesn't take up much room, and this allows me to use the extra space for something more wildlife friendly.

Plant hormones are produced at all stages of a plant's life and by nearly

Dwarf forms of plants common to horticulture, such as the dwarf Walter's viburnum (*Viburnum obovatum*) growing right of the standard form, are the result of a lack of hormones.

every part of its overall structure. A few, like auxin, have been known since the mid-1860s, but many others have been discovered in just the past few decades. As biologists recognize new plant hormones, we also discover how complex and intricate their various roles are. For the most part, plant hormones are produced in small concentrations, so clearly determining their individual roles is difficult. To compound this, they often work in unison with other hormones or in antagonistic relationships with other hormones. Teasing apart the role each hormone plays in these complicated relationships is not an easy task.

AUXINS

Auxin was the first plant hormone studied by plant biologists, and its influence has been recognized since the early Greeks. Charles Darwin and his son Francis published their research on plant movements in 1880, although the exact nature of auxin was unknown at the time. Their book *The Power of Movements in Plants* provided the foundation for the modern study of plant movements caused by hormones. The Darwins and their colleagues had long known that plants grow toward light. It is now understood that this response is controlled by auxin.

Indole-3-acetic acid (IAA) is the main auxin produced in higher plants, but plants produce other similar auxins such as Indole-3-butyric acid (IBA). Synthetic auxins also have been manufactured for use in rooting hormones and in some herbicides. All have profound effects on plant growth and development. Both plants, and some plant pathogens, can produce auxins to modulate plant growth.

Auxin stimulates differential growth in response to gravity. Such a response is critical to a plant's survival. Animals sense up from down using specialized organs not shared by plants. Plants sense gravity with their hormonal system and specialized cells called amyloplasts. Plant biologists call this sense gravitropism.

It is clear that plants can sense up from down. For the most part, their roots grow downward, and their shoots grow upward. We take this apparently simple response for granted, but the biological processes that control it are rather complex. Plant biologists are still working to fully understand it. Gravitropism requires the coordinated activity of different cell and tissue types. In plants, the area where gravity is sensed is often spatially distinct from the area of growth. Communication between these two discrete areas is necessary to coordinate everything effectively.

To date, gravity sensing in plants has been largely explained by the starch-statolith hypothesis. This hypothesis is based on the fact that the roots of most plants are known to contain gravity-sensing cells at their tips. These cells contain dense, starch-filled organelles known as amyloplasts, which settle to the bottom of the cells in response to gravity. The heavy amyloplasts then trigger the production of auxin. Higher levels of auxin are produced on the lower side of the root than on the upper side. This causes the lower side to elongate faster and turn the root downward in the direction of gravitational forces.

In a way, this is not so different from how humans sense gravity. Our sense of gravity is largely determined inside our inner ear. There are two

Plants can sense up from
down and grow accordingly.

chambers lined with sensory hairs embedded in a gelatinous matrix. On
top of this matrix are small crystals of calcium carbonate known as otoliths. These otoliths move as we change position, just like the starch gran-
ules in a plant's amyloplasts. The moving otoliths pull the hairs, and this
triggers a nervous system response in our brain. The miracle of plants is
that they have to make this adjustment without the benefit of a brain or
central nervous system.

The unresolved issue in plants is to understand how the physical movement and settling of amyloplasts in one set of cells triggers the accumulation of auxin in another, physically distant, set of cells. Current research has not been able to fully explain it. The most prevalent hypothesis is that the inner structure (the cytoskeleton) of each plant's cell plays a major role. It is possible that the proteins that comprise the cytoskeleton of the plant cell actually connect the gravity sensing cells to those that produce auxin. Regardless of the exact mechanism, however, it is clear that auxin is the major plant hormone involved in a plant's ability to send its shoot upward and its roots downward.

It is equally important that plants sense the direction of the sun and grow toward its light. This begins from the instant a seed germinates below ground. Newly germinated seeds have a limited amount of stored energy to reach the soil surface, generate leaves, and become photosynthetic. To accomplish all this before running out of energy requires seeds to be efficient. This initial growth upward toward the soil surface is accomplished by elongating their rapidly multiplying cells, and this is the result of auxin. Phototropins then guide the seedling in the shortest path possible to the sunlight.

The ability to sense sunlight and bend toward it is never lost in the mature plant. Auxin is produced at the tips of the shoots and passed from cell to cell using specialized proteins located at the base of each plant cell. Auxin is actively moved from the tip of the shoot to the cells below it without the influence of gravity. Export and import proteins, known as PINs, push the auxin out of one cell and into the next until the auxin eventually reaches its target site. This may seem counterintuitive, but auxins are transported below the tips, which causes these lower cells to elongate, pushing the stem upward. This same process is what causes stems to realign themselves if the plant is turned sideways. Auxins are then concentrated more to the lower portion of the now-sideways stem, which causes the lower portion to elongate faster than the upper surface, and the unequal elongation of cells causes the stem to turn upward once again. The light response obviously requires a complex system of light receptors within the plant that work together to cause the target plant cells to produce this hormone and different transport proteins that work to distribute it to the proper location. Sensing the direction of the sun and growing toward it is called phototropism.

Growing upward toward the light is sometimes referred to as the "shade-

The hormone auxin does a great many things, including reorienting a plant if it falls over.

avoidance effect." Plants initially subjected to the shade of the understory typically put most of their energy into growing upward, often to the exclusion of gaining girth, until higher light levels are achieved. This stretching skyward occurs under the influence of auxin.

Auxin does much more than guide the stems toward light and the roots downward. It is the primary hormone responsible for maintaining apical dominance—directing the plant to only grow at the tips of the stems. Without auxin, plant growth would be completely haphazard. Buds along the side of the stem would swell and start to grow, and the plant would then produce shoots in all directions. The auxin produced at the tips of the main stem and branches prevents this type of growth and keeps the axillary buds from developing. It is in the plant's best interest to focus growth at the ends of its main stems.

The influence of auxin in inhibiting the growth of axillary buds is obvious if the ends of the main stem are damaged or removed by pruning. With the primary growing points removed, the side buds rapidly are released from the inhibitory influence of the auxin being produced there, they quickly swell, and new growth is initiated. Eventually, new stems take on the role

Auxin also is involved when a plant loses its growing tip and once-dormant buds are released to take over that role.

that the old ones performed, and they become the new primary sites for growth, but until the dust settles, a great many axillary buds may make a bid for that job, and the plant becomes noticeably bushier. Frequent pruning causes chaos with this process and does not allow for apical dominance to fully be restored; all the side branches keep growing nearly equally, and the plant becomes shorter and bushier.

At times, this is preferable in a landscape, but at other times it needs to be corrected. My turkey oak (*Quercus laevis*) suffered extensive damage to its main stem shortly after it was planted. For the next several years, several of its side branches vied for the role of the apically dominant stem, and it became, for all intents and purposes, a rounded shrub. I corrected this by choosing one of the stems and pruning away all the others. The remaining stem, without the competitors, quickly asserted its sole apical dominance, and it is a beautiful single-trunked tree today.

Auxin is the plant hormone responsible for lateral root formation in cuttings. As such, it revolutionized the horticultural industry when synthetic root-stimulating compounds became available. If you have ever used a root-stimulating powder to root a cutting from one of your plants, you have used a synthetic auxin.

Auxin does the same thing in intact plants. The formation of lateral roots is responsible for the architecture necessary to stabilize a plant in the ground. While a taproot initially holds a seedling firmly in the ground, it's the development of the lateral root system that stabilizes it and forms the system most responsible for water and nutrient uptake. Auxin is the primary plant hormone responsible for the formation of the lateral root system. Research studies have shown that auxin transport within the plant stimulates the development of lateral roots from within the pericycle—the inner cell layer of the root responsible for lateral root development. Though the exact mechanism involved is still a mystery, new lateral roots do not develop from within the pericycle without the presence of auxin.

The formation of new leaves in the apical meristem is initiated by the accumulation of auxin. Research studies have shown that auxin is the hormone responsible for every stage of leaf formation and development including leaf shape, cell differentiation within the leaf, and the spacing of leaves along the stem. Since the mid-1930s, physiological experiments have shown that leaf formation at the stem apical meristem is dependent on the delivery of auxin through the cells via the PIN proteins described above. The convergence of auxin flow from epidermal cells at the tips of the shoots creates an accumulation of auxin activity at the edge of the shoot apical meristem, and these sites mark the points where leaves are then initiated.

While this is occurring, PIN proteins in cells below the epidermis of the newly developing leaves direct auxin flow through the center of the leaf and define the position of the future midvein; midvein formation is based on directional auxin transport.

Auxin is also the hormone responsible for the spacing between leaves along the stem. Already-developed leaves deplete the surrounding cells of auxin so that new leaves do not form too close to them. Most plants maintain a characteristic distance between leaves that is modified to some extent by sunlight and other growing conditions. The general distance, however, is largely determined by auxin.

A key event in the development of leaves is the correct differentiation of the specialized cell types that underpin the physiological functions of the leaf, such as stomata for gas exchange, vascular cells for the transport of water and nutrients across the leaf, and mesophyll cells for photosynthesis. Although the cellular proliferation, differentiation, and expansion stages of leaf development cannot be rigidly separated in time, research suggests that the temporal regulation of the transition from one stage to another is dependent on auxin.

Deciduous plants lose their leaves at well-defined seasons during the year, particularly plants subject to winter cold or distinct wet and dry seasons. Leaf loss during these times is a marked ecological advantage if the opportunity to photosynthesize and grow is far overshadowed by the need to conserve water and wait for better times. The loss of leaves, known as leaf abscission, is a significant event for deciduous plants, and it is controlled through auxin and several other plant hormones that will be discussed below. Research has shown that leaf abscission is actually a two-stage process. Auxin has been shown to inhibit leaf abscission during the first stage as the leaf is initially preparing to separate from the stem, but it accelerates abscission during the second stage once these preparations are made. Though several theories have been proposed by research scientists as to how this works, it seems clear that the timing of the arrival of auxin to the abscission zone plays the major role in the overall process.

Leaf drop is a complicated process that involves a number of plant hormones.

For nearly a century, auxin has been recognized for its effects on post-embryonic plant growth. Recent insights into the molecular mechanism of auxin transport and signaling are uncovering fundamental roles for auxin in the earliest stages of plant development, such as in the development of the apical-basal (shoot-root) axis in the embryo, as well as in the formation of the root and shoot apical meristems and the cotyledons. Localized surges in auxin within the embryo occur through a sophisticated transcellular transport pathway. The resulting downstream gene activation, together with other, less well-understood regulatory pathways, establish much of the basic body plan of the plant embryo.

Plant biologists have only recently unraveled the role that auxin plays in the development of cell and organ types in the plant embryo. This is partly because the plant embryo is safely, and for many experimental purposes, inaccessibly, tucked away inside the ovule. It is also because they have had, until recently, a very limited understanding of the molecular mechanisms that underlie auxin action. Despite these difficulties, the profound effects of auxin on in-vitro-cultured plant tissues have suggested that auxin has the power not just to regulate growth but also to dictate the fate of a cell. Research experiments, for example, showed that the ratio of auxin to cytokinin (another plant hormone) in the growth medium determined whether roots or shoots developed from cultured cell clusters. The concentration of auxin in the medium could also be manipulated to cause cultured mesophyll cells to develop into xylem cells. Additional research has clearly demonstrated that auxin is significant in the overall development of the plant embryo and that it directs a great many things including the establishment of the root and the partitioning of the shoot apical meristem into the cotyledons and the shoot apical meristem itself. As the embryo's meristematic tissues generate new cells, they are largely undifferentiated. The presence of auxin is necessary for determining which type of cell they will mature into.

Finally, auxin plays a role in fruit development. Though fruit development is a complex process involving many plant hormones, active fruit growth is largely due to the synthesis of auxin in the developing seeds. Synthetic auxins, applied to the unfertilized ovary of flowering plants, induce fruit development. These auxins are blocked at the base of the flower under normal development until fertilization has occurred, but once this blockage is removed, it is the transport of auxins into this region that is primarily responsible for the development of the fruit. The hormonal regulation of ripening has also been shown to be dependent on the active involvement of auxin in its influence on the synthesis of another plant hormone, ethylene.

GIBBERELLINS

During the 1930s, Japanese plant biologists isolated a growth-promoting substance from rice parasitized by a fungal pathogen that caused the rice to grow erratically, an effect known as "foolish seedling syndrome." The substance responsible, gibberellic acid (gibberellin), was found to actually occur naturally in plants. The fungal pathogen was not directly responsible for the rapid rise of gibberellic acid in this infected rice; it stimulated the rice indirectly to produce it itself.

Gibberellic acid has a number of effects on plant growth, but the most dramatic is its effect on stem growth. Unlike auxin that causes cell elongation, the presence of gibberellic acid causes stems to grow rapidly. When applied to "bush-type" beans, for example, they become indistinguishable from climbing or "pole" beans. Gibberellin also induces stem growth in rosette plants such as cabbage (*Brassica oleracea*) and broccoli (*Brassica oleracea* var. *italica*). Rosette plants have profuse leaf growth and retarded stem growth. Just prior to flowering, however, the stems elongate enormously, a process known as bolting. Bolting requires long days or cold nights to occur naturally. Under these conditions, the plants begin producing increased amounts of gibberellic acid. When a cabbage head is kept under warm nights, it retains its rosette habit. Bolting can be induced artificially by the application of gibberellins to the tip of the stem.

On the other side of the spectrum, many dwarf or compact forms of plants sold in the horticultural trade or grown as food crops are deficient in gibberellic acid or have genes that do not recognize its presence. The introduction of dwarfing genes into wheat, for example, was a major factor in breeding higher-yielding varieties during the "green revolution" of the 1960s, because more nitrogen fertilizer could be applied without leading to excessive stem growth and having the ripened seed heads fall over—a process known as "lodging." Such varieties are still widely grown commercially. The importance of dwarfism has also been applied to other major grain crops. Gibberellic acid–insensitive dwarf mutants in corn, rice, and barley have all made significant contributions to modern agriculture.

Gibberellic acid is involved in a great many other plant functions than stem growth. This hormone is required by the embryo while still in the seed for it to develop and germinate. Seeds that accumulate gibberellic acid prematurely may actually sprout while still inside the fruit. The balance of gibberellic acid inside the maturing seed is critical to both the development of the embryo and the correct timing of germination. Throughout the process, there is a precise "dance" between it and the hormone that retards its production while the embryo grows—abscisic acid. Maturing embryos begin producing gibberellic acid, and this counteracts the inhibitory influences of abscisic acid. Ultimately, this leads to germination.

These same two hormones are involved in bud dormancy. Dormant buds can be induced to sprout by treating them with gibberellic acid, although it is not the only hormone involved. In temperate climates, buds remain dormant throughout the winter and actively expand and bolt in spring. Gibber-

Gibberellic acid is an important plant hormone responsible for many aspects of plant growth. Bolting in many common vegetables, such as those in the cabbage (*Brassica oleracea*) family, is largely the result of gibberellins.

Hormones also prevent seeds from sprouting prematurely. A hormone imbalance in this tomato (*Solanum lycopersicum*) has allowed these seeds to sprout while still inside the ripening fruit. Photograph by Juliet Rynear.

Gibberellic acid, along with abscisic acid, are involved with bud dormancy. This interplay of hormones keeps the buds from swelling before conditions are right and allows the buds to expand when they are.

ellic acid synthesis takes place inside the plastids, alongside photosynthesis. As such, its production increases as day length increases. As spring activates new growth and the initiation of photosynthesis, gibberellins are released from the plastids into the cytoplasm of the bud cells, and this activates the buds to begin growing.

Finally, gibberellins influence flower production and fruit size. Pollination stimulates the production of certain forms of gibberellic acid within the flower's ovary, and this helps to ensure fruit set. In fact, flowers treated with gibberellic acid may produce fruit even if they are not pollinated. Such fruit are obviously seedless, and this phenomenon has been used in modern agriculture. Orchardists involved with the production of stone fruit crops, such as peaches and plums, as well as apples and pears, have also taken advantage of gibberellins. When such trees are treated with foliar applications of gibberellins in early spring, they produce fewer flowers but larger fruit that are less likely to be aborted before they ripen. In nature, the production of gibberellins helps to ensure that plants do not produce more fruit than they are capable of bringing to full term.

CYTOKININS

Cytokinins are involved in cell division and the making of new plant cells and organs. As such, they are absolutely critical to the daily life of plants. Cytokinins are produced in the tips of the roots (the cells of the apical meristem) and then are transferred upward throughout the plant dissolved in the water and nutrients within the xylem transport system. As plants transpire during the day, the one-way movement of water upward facilitates the distribution of cytokinins. It is an elegant system that requires absolutely no energy to function.

The name cytokinin is derived from its ability to induce cytokinesis, the physical process of cell division. Though cytokinins stimulate the process, the ultimate effect is controlled by auxins. When cytokinin is provided to a culture medium containing plant cells, the protein-synthesizing activity of the cells is greatly stimulated. Moreover, some of the synthesized proteins are new ones. Such observations suggest that cytokinins control cytokinesis by regulating the synthesis of specific protein factors that are required for cytokinesis to occur.

Cytokinins are important in the rapidly growing business of producing plants through tissue culture, a process that has revolutionized the horticultural trade. Consider orchids, for example. When I was a child, orchids were produced by seed or by the division of mature plants. Divisions are extremely inefficient at producing large numbers of plants for sale, and growing orchids by seed is extremely slow and fraught with a great many challenges in getting the seedlings to marketable size. Quality control was also difficult because seed-grown plants do not always share the superior attributes of their parents. Keeping orchids was a costly hobby.

Modern tissue culture, however, changed all that. In 1962, plant biologists developed the first really reliable tissue culture medium. Plant cells could then be taken from superior plants, separated out across a growing plate and caused to become new plants. Plant cells taken from actively growing portions of the plant are undifferentiated—they can become any type of cell possible. Each of these cells, therefore, can be made to start growing and dividing and eventually each can become a whole plant that is an exact clone of its parent. Award-winning parents can be cloned using tissue culture to produce thousands of exact copies in far less time than it

Tissue culture, using the plant hormone cytokinin, has allowed for the mass production of many plants, such as these orchids, that were once time-consuming and difficult to produce.

would take to grow a fraction of this number by seeds, and the quality of each is assured. Thus, you can now find quality orchids at grocery stores, discount box stores, and roadside stands for a fraction of what a good plant once sold for during my childhood. Though auxins, nutrients, vitamins, and other additives are important to a tissue culture medium, it was the discovery and isolation of cytokinins in 1953 that made this revolution possible.

Cytokinins play a number of other significant roles besides the production of new cells and plant organs. They are involved in cell enlargement, fruit development, and growth of the stems and leaves. Cytokinin-deficient plants develop stunted shoots with smaller apical meristems and significantly smaller leaves. In contrast, the root meristems of cytokinin-deficient plants become enlarged and give rise to faster-growing and more branched roots. This suggests that cytokinins are an important regulatory factor of plant meristem activity and cell production, with opposing roles in shoots and roots.

The cotyledons of some plants expand dramatically when they are treated with cytokinin. This result is due to cell enlargement rather than cell division. During cytokinin-induced cell enlargement, respiratory activity increases significantly and greater amounts of potassium ions accumulate in the cells. This suggests that cytokinins bring about permeability changes within the cell membranes.

Cytokinins are the fountain of youth in plants as they delay the natural aging process that leads to plants' death. Cytokinins encourage even older cells to divide. As long as cytokinins are being actively produced, those cells tend to remain active and healthy. When production stops, those cells quickly age and die. It is one of the most important plant hormones, for example, in leaf senescence. Without cytokinins, there would be no reason for road trips to see good fall color.

Cytokinins are involved in repair, too. If a plant becomes wounded, it can fix itself with the help of cytokinins and auxin. If the concentration of auxin and cytokinins is equal, normal cell division will take place. If the concentration of auxin is greater than that of cytokinins, roots will form—the theory behind root-stimulating powders. If the concentration of auxin is less than that of cytokinins, however, shoot production will exceed root production.

Last, there is developing evidence that cytokinins play a role in controlling certain plant diseases. Auxin production can actually increase a plant's susceptibility to disease and pathogen attack, and some pathogens stimulate their host to produce more auxins after infection to create conditions favorable to them. Cytokinins tend to have a counterbalancing effect on many plant responses, and recent research suggests that cytokinins counteract the influences of these types of infections.

Science is only recently unraveling how plants actively fight disease and pests. Cytokinins can be instrumental in mediating host susceptibility to fungal pathogens by walling off this infection and generating a green island around it. High levels of cytokinins also increase the resistance of plants to some viral pathogens and herbivores. Even certain soil microbes surround-

Fruit ripening is stimulated by the production of the plant hormone ethylene.

jor factor in the senescence of leaves and their eventual fall. Even the leaves of evergreen plants don't last forever, and plants need to prepare for their loss. As discussed above, plants initiate the process by first weakening the connection between the leaf and stem and then forming a scab to seal this area. Ethylene is the hormone primarily responsible for inducing the production of various enzymes that break this layer down prior to leaf fall. It does the same for certain fruits like apples and pears that drop to the ground when fully ripe.

As discussed above, apical dominance is conferred by auxin, and lateral branch growth comes when cytokinins are given temporary dominance following damage or loss to the apical buds. Ethylene also has a part to play in all of this. Recent research has demonstrated that as auxin levels increase in the apical stems, the production of ethylene is stimulated and then diffused into the lateral buds. The elevated levels of ethylene, more than the presence of auxins, are responsible for blocking the development of the lateral branches. Auxins work hand in hand with ethylene: auxins stimulate the production of ethylene, and ethylene inhibits the transport of auxin to the stem tips. The same phenomenon seems to be responsible for gravitropism in roots. Though auxin is the hormone ultimately responsible for the ability of roots to detect the direction of gravity, auxin-induced production of ethylene influences the overall relationship. Ethylene-treated roots lose their sense of directional growth.

Ethylene influences the formation of flowers and their eventual senescence. It also influences seed germination in some species, stimulates stem growth in underwater species, and stimulates the growth of root hairs. As with so many plant hormones, ethylene has a great many roles to play.

ABSCISIC ACID

Abscisic acid is the last of the so-called major plant hormones. Discovered in the early 1960s, it regulates many aspects of plant growth and development, including embryo maturation, seed dormancy and germination, cell division and elongation, flower production, and responses to environmental stresses such as drought and pathogen attack. Abscisic acid is found throughout the plant kingdom and is synthesized indirectly from the carotenoid pigments that assist in photosynthesis. Unlike its name suggests, however, it does not direct control of leaf abscission. That is the work of ethylene and auxins.

What abscisic acid may be best known for is its role in plant stress, especially drought. When confronted by stress, animals produce cortisone-based hormones such as adrenaline. Plants produce abscisic acid. Unlike animals, plants have to cope with the conditions with which they are confronted. They cannot run away or migrate to more favorable locations. What looks on the surface to be more simplistic actually makes plants more complex. They have to be masters of whatever location they take root in. When faced with drought conditions, plants must quickly counteract the loss of water by closing their stomata. They do this by sending abscisic acid to the guard cells responsible for keeping the stomata open. In the presence of this hormone, water leaves the guard cells and the stomata close down. For as long as the plant continues to send abscisic acid to the guard cells, water loss through the stomata is prevented. The use of abscisic acid in commercial agriculture helps prevent the impacts of projected drought even before it has become a serious problem. Foliar sprays of this hormone cause an almost immediate effect on crops and help the plants conserve water before the natural production of abscisic acid would occur. Of course, closing the sto-

The plant hormone abscisic acid is involved in a great many functions, including retarding wilting.

mata also essentially stops photosynthesis and the ability of plants to grow. It is a trade-off—temporary loss for future gains.

Water stress is not the only stress faced by plants. Wounded plants produce abscisic acid, as do those faced with salts and cold temperatures. Faced with these kinds of stresses, plants need to alter their normal growth and development to mitigate the potential damage. Abscisic acid is involved in regulating these adaptive responses until such time as the dangers have passed. The production of abscisic acid has clearly shown to increase a plant's ability to survive cold temperatures, attacks by insects, and increasing salt levels. It is a plant's best coping mechanism when times get tough.

Abscisic acid also inhibits the stimulatory effects of other plant hormones. It is a plant's calming mechanism as well as its coping one. For example, whereas gibberellins can induce seeds to start breaking down starches prior to germination, abscisic acid works to delay it. Seed dormancy and the production of proteins within seeds are the result of abscisic acid.

OTHER PLANT HORMONES

Though most botany texts and online sources focus on the five so-called major plant hormones discussed above, much of the cutting-edge work on plant hormones is based on other hormones discovered in the past few decades and for which a developing body of information is now emerging. While there is still much that isn't known, what is clearly evident is that plants are every bit as complex as animals in terms of their hormonal activity. Hormones help regulate every possible activity of a plant's daily life, many occur in extremely small amounts, and they almost always work in unison with or opposite to other hormones to create a complex balance that fine-tunes their responses to the world around them.

Plant Defense Hormones

Because plants cannot avoid danger by moving away, they must form defenses against pathogens, competitors, and herbivores while standing their ground. In nature, there is no rolling over without a fight even if the fight is going on sight unseen. Over the past few decades, a growing body of information has developed around plant defense mechanisms and the hormones that plants produce to induce them. Just as animals produce antibodies in response to foreign antigens, plants produce hormones in low concentrations that stimulate the production of "antibody" chemicals.

One of these hormones is salicylic acid, the basic ingredient of aspirin. Though isolated from willows, salicylic acid operates as a hormone in plants by activating genes that cause plants to produce chemicals that ward off pathogenic invaders. Salicylic acid not only plays a role in the plant's production of proteins designed to fight a pathogen at the invasion site, but its presence in the phloem induces resistance for the rest of the plant. It has been demonstrated that infected plants convert salicylic acid to a volatile organic compound (an ester) and release it into the air through their stomata. Nearby plants pick up these esters through their open stomata and prepare their own defenses against possible infection. Salicylic acid tends to be most effective on pathogens that require living hosts to complete their life history. Plants use different hormones for pathogens meant to kill them outright.

While the salicylic acid pathway is critical for these types of responses, it is not the only signaling pathway mediating plant defenses against patho-

Herbivory causes plants to produce several hormones that reduce their palatability and/or level of nutrition.

gens. To build an optimal and efficient defense, plants have developed multiple pathways, and they are capable of cross-talking between these pathways to fine-tune and specify their responses.

Jasmonates are plant defense hormones produced from fatty acids. Unlike salicylic acid, however, they are most effective against pathogens meant to kill them and against herbivorous insects such as grasshoppers. As such, jasmonates (as well as ethylene) provide a first line of defense for a great many potential pathogens and herbivores.

Although jasmonates regulate many different processes, their role in wound response is best understood. Following mechanical wounding or herbivory, plants quickly activate jasmonates, which trigger responses from appropriate genes. For example, when tomato plants are being fed on by tomato hornworm caterpillars, the plants quickly begin producing defense molecules that inhibit leaf digestion in the caterpillar's gut. Another indirect result of jasmonate signaling is the emission of various volatile organic compounds through the leaves that then travel airborne to nearby plants and cause them to also respond by increasing their levels of jasmonates.

Wound and pathogen responses seem to be negatively related. For example, reducing a plant's production of salicylic acid lowers its ability to respond to various pathogens but enhances the production of jasmonates and the plant's ability to resist herbivorous insects. Generally, it has been found that pathogens living in live plant cells are more sensitive to salicylic acid–induced defenses, whereas herbivorous insects and pathogens that derive benefit from cell death are more susceptible to jasmonate defenses. Thus, this trade-off in pathways optimizes defense and saves plant resources.

Jasmonates, salicylic acid, and ethylene are not the only plant hormones involved in plant defense. In 1962, the first polypeptide plant hormone, systemin, was isolated in tomato leaves and shown to be involved in protecting the plant against herbivory. Since then, related polypeptide plant hormones have been isolated in a variety of plant species. Leaf damage by herbivores or by mechanical wounding induces the activation of at least fifteen defense genes throughout the entire plant within hours. The primary chemical involved in activating these genes is systemin. Systemin is found in the cytoplasm of plant cells and within the cell walls themselves. It is released rapidly when the cells are damaged and readily transported from wound sites throughout the plant. Plants exposed to systemin quickly synthesize a complex elixir of defense proteins that, among other things, reduce the nutritional value of the plant to future herbivory. Research has demonstrated that the production of systemin causes a significant decrease in the damage to leaves caused by herbivores. The continuous activation of this hormone, however, is costly to the plant because it affects its growth, physiology, and reproductive success. There is a trade-off. When systemin production was prevented in experimental settings, larvae feeding on the plants grew three times faster than larvae feeding on normal plants, but much of the new growth was directed to the roots and not the stems. This defensive mechanism helps to ensure that the damaged plants will have sufficient energy stored up below ground to recover more quickly once the herbivory has ended.

As with other plant hormones involved with defense, systemin plays a critical role in defense signaling, communicating critical information between plants via the release of volatile organic compounds into the air. Therefore, entire populations can begin protecting themselves prior to an attack. There also is evidence that different herbivores produce different chemical responses in the plant. Plants can detect the difference between rabbit herbivory, for example, and the "herbivory" caused by a pruning shear, and their production of various protein inhibitors is modified appropriately as well as the types of volatile compounds released into the atmosphere. In tomato and in wild tobacco (*Nicotiana* spp.) plants, for example, systemin-caused release of volatile organic compounds attracted specific wasps to the plants that then parasitized the caterpillars feeding on them.

Plants live in complex environments in which they intimately interact with a broad range of other organisms. Besides the many potentially harmful interactions with pathogens and herbivores, relationships with beneficial microorganisms are frequent in nature as well, improving plant growth or helping the plant to overcome stress. The evolutionary arms race between plants and their enemies has provided plants with a highly sophisticated defense system that, as in animals, recognizes molecules that are not its own or signals from injured cells, and responds by activating an effective immune response against the invader encountered.

Recent advances in plant immunity research underpin the pivotal role of cross-communicating hormones in the regulation of the plant's defense signaling network. The hormones' powerful regulatory potential allows the plant to quickly adapt to its hostile environment and to utilize its resources in a cost-efficient manner. Plant enemies, on the other hand, can hijack the plant's defense signaling network for their own benefit by affecting hormone homeostasis to antagonize the host immune response. Similarly, beneficial microbes actively interfere with hormone-regulated immune responses to avoid being recognized as an alien organism. In nature, plants simultaneously or sequentially interact with multiple beneficial and antagonistic organisms with very different lifestyles. However, knowledge on how the hormone-regulated plant immune signaling network functions during multispecies interactions is still in its infancy.

Another area of plant defense and hormone production that is still poorly known is how environmental factors influence this regulatory network. Light signaling has been shown to strongly influence hormone-regulated defenses. Plants seem to display an altered sensitivity to jasmonate-dependent defenses under conditions of crowding and shade. This results in a trade-off between resource investment in defense and increased growth to outcompete neighbors. Cross-talk between abscisic acid–mediated drought stress and jasmonate-dependent defense against pathogens was recently shown to be regulated by hormonal-produced enzymes, and both reactions were shown to be influenced by the production of another plant hormone, salicylic acid. In other situations, however, otherwise antagonistic hormones seem to work together. In heat-stressed plants, salicylic acid and jasmonates seem to act together to reduce this stress. All this highlights the complexity of the regulatory processes involved in the adaptive response of plants to their continuously changing environment.

Like all defenses, this complex system of plant hormones is vulnerable to hacking by its enemies. Fusarium fungi that cause wilting and death in some plants, for example, has been shown to promote disease by taking control of the jasmonate pathway and rendering it useless. Insects have been demonstrated to exert similar tactics. For instance, nymphs of the phloem-feeding silverleaf whitefly were shown to induce the salicylic acid pathway while suppressing the jasmonate one, resulting in a faster development of the nymphs. Caterpillars of the beet armyworm cause similar disruptions to the plant's defenses through their saliva. Some insect eggs, oviposited on certain plant tissues, activate the salicylic acid pathway and thereby suppress the jasmonate-dependent one, allowing the newly hatched larvae to feed on virtually unprotected foliage.

Diseases and insect damage cause plants to secrete a different set of hormones than when they are under attack by herbivores. These hormones, like the antibodies in our own systems, work to fight off the effects of these pathogens. This plant is being attacked by white flies. Photo by Elizabeth Hoefler-Boing.

Things can get even more complex at the root zone as plant roots interact with an incredibly diverse community of soil microbes and invertebrates that can also influence a plant's overall defense status. Below-ground problems can also result in affecting the above-ground hormonal signature and lead to certain levels of resistance against herbivores and pathogens.

Below-ground interactions are frequently beneficial in nature. Because beneficial microbes are also recognized as alien organisms, active interference with the plant immune signaling network is fundamental for the establishment of intimate mutualistic relationships. In many cases, salicylic acid, jasmonates, and ethylene were found to be the dominant regulators in this process. Suppression of ethylene signaling has been observed in tomato plants with beneficial mycorrhizal fungi relationships in their roots. In addition, plant growth–promoting bacteria have been shown to actively suppress ethylene-dependent microbe-associated responses in the roots of cress plant, presumably to avoid being recognized as an uninvited guest. Colonization of roots by beneficial rhizobacteria induces a systemic resistance in aerial plant parts that is effective against a broad spectrum of attackers. The list of examples by which beneficial microbes recruit hormone signaling pathways to establish a mutualistic interaction is growing rapidly. However, our understanding of how plants regulate their immune signaling network to maximize both profitable and protective functions during simultaneous interactions with the good, the bad, and the ugly is still limited.

Growth Regulators

All plant hormones have some impact on plant growth and development, but a few besides the "major five" play key roles in these functions. Brassinosteroids are a relatively recently discovered group of plant hormones responsible for regulating plant growth. First described in mustards, brassinosteroids have been shown to stimulate cell elongation and division, gravitropism, resistance to stress, and xylem differentiation. They also inhibit root growth and leaf abscission in conjunction with other plant hormones. Plants that lack brassinosteroids remain small and are often male sterile.

Polyamines are found in animals and plants. In plants, they are essential for normal growth and development, and they affect cell division. Strigolactones have been implicated in the inhibition of shoot branching, along with auxins.

Stress Hormones

Plants feel stress in their daily lives just as animals do, and not all of it is related to pathogens and herbivores. Abiotic environmental stresses are brought on by such things as drought, salt, and nutrient deficiencies. To cope with these types of stresses, plants produce a cocktail of hormones that allow them to deal with each unique situation. Recent research has shown that plant responses are both complex and fine-tuned to each individual stress. Although abscisic acid has long been identified as the primary hormone involved with environmental stress, a great many others have been recently implicated. Nitric oxide signals hormonal and defense responses while regulating the synthesis of energy (ATP) within plant cells, root development, seed germination, cell death, and stomatal closing. Jasmonates and salicylic acid have been shown to assist plants with dealing with water and salt stress. Systemins also are important. Systemins are sometimes referred to as "conditioning" hormones as their presence prepares plants to deal with environmental stressors such as prolonged drought or salt spray. As in animals, polypeptide-based hormones play important roles in mediating communication between plant cells.

Flowering Hormones

The plant hormone primarily responsible for inducing flowering, florigen, was not isolated and described until 2005, though it was suspected for nearly one hundred years prior to that. It was, and still largely remains, an elusive hormone.

Florigen is just one of possibly many plant hormones involved with the process of flowering.

Florigen is produced in the leaves and affects the growing tip. Here, it acts like a molecular alarm clock, wakening a complex set of reactions that ultimately leads to flower formation. Of course, florigen is not produced in sexually immature plants not yet old enough to reproduce, but it is triggered in mature plants based on their detection of day length and other environmental factors.

Though still poorly understood, florigen could someday be used to manipulate plants to flower on demand. Instead of having a discrete blooming season, horticulturally produced specimens might be made to flower out of their traditional blooming times and/or made to flower sequentially for extended periods. Although this would be a boon to the commercial trade, there are good ecological reasons why plants do not do this naturally. The long-term health of a plant would likely be jeopardized by forcing it to flower for prolonged periods without a rest or in months its pollination biology has not evolved to coincide with.

11

Plant Communication

James Cameron's 2009 movie *Avatar* is a visually spellbinding epic set in a world populated by a native people physically and spiritually in touch with the world around them and interconnected by the Tree of Souls that acts as a supercomputer capable of storing and uplinking the society's collective knowledge. This seems to be pure fantasy, but Cameron must have had some knowledge of plant biology to have imagined this model because it closely mimics what we now know about plant communication. In fact, the Tree of Life model portrayed in *Avatar* is actually less complex than the reality of plant biology; communication in the plant world is much more complex than anything we would have guessed at just a few years ago. We are just now beginning to unravel that complexity.

This should not surprise us, however. Plants are complex living organisms, and communication is vital to every aspect of life on earth. Animals communicate their feelings and intentions in verbal ways and in a great many nonverbal ones. Songbirds, for example, may use a dozen or more calls and songs to convey information to the world around them. Human birders can distinguish these various calls and ascribe individual behaviors to each of them. Territorial calls are different than alarm calls, for example. We "talk" to our dogs and generally learn to understand their vocalizations as well. Animals of all kinds use body language as well as vocalizations. It often is possible to read the body language of our partners without sharing a word, and heaven help us if we err. Most animals' ability to read body language far exceeds ours. Their survival depends on it.

Evolution favors those of us who can communicate effectively—communicating our thoughts and needs as well as listening to the communications of others. It only makes evolutionary and ecological sense that

Plants, like animals, have evolved the need to communicate with each other. They just do it differently. Photograph by Christina Evans.

plants have developed their own system of communication. Though we have been largely left out of the "chatter" generated by plants, we should not make the mistake of considering them primitive or one-dimensional. Everything an animal feels compelled to communicate is equally important to plants. Their long-term survival depends on their having developed multiple methods to communicate with one another.

Just as animals have developed multiple methods to communicate to a diverse biological audience, plants have evolved to do much the same. Methods that might work within a plant species are often modified so that they can communicate with other species, and modified again for soil microbes and microorganisms necessary for the life of that plant. Plants have a lot to talk about, and they do it in very diverse ways, but if you think about it, that only makes sense. We are just now beginning to unravel these complex systems.

ABOVE-GROUND METHODS

Plants communicate with each other above ground by releasing scents into the air. Some of these are scents that we can detect too, but a great many are detectable only by other plants or by animals that the plants specifically target. Biologists call these scents "volatile organic compounds," VOCs for short. When we mow our lawn and sense that smell of freshly cut grass, we are picking up the lawn's alarm call telling the plants around them that danger is nearby. Likewise, the scent of a flower or the crushed leaf of an herb are VOCs, released for a purpose that provides plants an ecological edge. We know that plants release VOCs. What we are now learning is that these volatile chemicals are part of a complex communication system used by plants to elicit responses in other plants.

One function of releasing VOCs is to increase the resistance of neighboring plants to the attack of herbivores. This function has been well documented by a variety of research conducted over the past three decades. In 1983, researchers demonstrated that Sitka willow (*Salix sitchensis*) growing close to herbivore-infested willows exhibited higher levels of resistance to herbivores than did plants growing farther away. The willows under herbivore attack seemed to be protecting neighboring willows. Even more surprising was that undamaged poplar (*Populus* × *euroamericana*) and sugar maple (*Acer saccharum*) saplings also increased their antiherbivore defense when exposed to the air around the damaged willows. Plants of several species were seemingly in communication with one another. Though some biologists initially questioned this idea, numerous experiments over the past few decades with dozens of other plant species have shown the same types of responses.

Although all plants release chemical scents into the air regardless of their status, they release far more chemicals and a wider variety of them when they are attacked. Wheat seedlings without damage attract aphids, but once they are under attack they release VOCs that repel other aphids. In this system, they regulate the amount of damage to a level they can seemingly endure. Research also has shown that host plants release VOCs that enhance or reduce the amount of egg laying that certain butterflies perform on them. When a grasshopper, a rabbit, a moth or butterfly caterpillar, or any other herbivore begins chewing on a plant, the injured plant fights back by altering its internal chemistry to make it less palatable and by releasing airborne chemicals to warn its neighbors to do the same in preparation for a possible future attack. Plants under attack by herbivores

The presence of aphids on a plant causes a variety of above- and below-ground communication methods to be stimulated.

also can produce specific chemicals in the herbivore's gut that get converted into signaling chemicals that attract the herbivore's predators. This type of system is rather devious. It is not passive; the specific chemicals produced by the plant to initiate this process are produced only once the plant is under attack by specific types of herbivores.

This is difficult to study in nature because plants are constantly releasing VOCs, and discerning their responses to each is complicated, but in controlled experiments lima beans (*Phaseolus lunatus*) exposed to herbivore-induced VOCs lost less leaf area to herbivores than those grown in air where the VOCs were filtered out. Plants exposed to the VOCs of their neighbors can "prime" themselves before they are attacked. Studies on a wide variety of species, including corn and lima beans, have shown that they suffered less damage and recovered more quickly when exposed to the VOCs produced by their neighbors. When given a heads-up, plants can prepare themselves for the onslaught that is yet to come.

Since 1859, when Charles Darwin published his now-famous *The Origin of Species*, biologists have argued about whether there is an evolutionary advantage for altruism or whether all the actions of living species are really done for their own best interests. Sending signals to neighboring plants might seem to increase the fitness of their neighbors without improving their own, resulting in an evolutionary disadvantage for the emitter. Because VOCs have a limited range before their signals are too dilute to be understood, perhaps such a strategy evolved to protect plants that were likely relatives of the one under attack. Research, however, has shown that completely unrelated species are capable of "eavesdropping" on these signals and increasing their fitness too. The evolutionary explanation for this phenomenon is problematic as this kind of altruism does not make sense on the surface, but it occurs nevertheless.

The emitters of VOCs can benefit themselves when their airborne chemicals attract predators that feed on their herbivore attackers. This type of response seems a bit far-fetched at first glance, but it has been demonstrated many times in field studies of various unrelated plant species. Parasitic wasps and carnivorous mites, for example, can cue in on specific VOCs released by plants and use them to zero in on their prey. Using VOCs, these insect predators discern infested plants and ignore those that aren't.

Some research has shown that plants emit different VOCs depending on which type of herbivore is attacking them. To do this, plants have to recognize their specific attacker and then emit a species-specific VOC that will alert a specific predator. Such complex signaling must take a great many years to evolve, but plants have been on land with insects for millions of years. It is a dance that is constantly evolving.

Of major importance to this relationship is the plant's ability to recognize the herbivore that is attacking it. It does this by analyzing the saliva secreted by the herbivore while it is feeding. In this way, plants can differentiate the general wounding caused by a pruning shears or lawn mower from that of a herbivore bent on chewing it up. VOCs are energetically expensive to produce. It does the plant little good to waste a lot of energy fighting an enemy like a hedge clipper.

Volatile organic compounds also have been shown to trigger herbivore resistance in the undamaged tissues of the plant under attack, not just in its neighbors. Plants seemingly perceive their "damaged self," as Martin Hull writes in a 2009 opinion piece ("Damaged-Self Recognition in Plant Herbivore Defence," *Trends in Plant Science* 14, no. 7) and thereby maintain the upper hand in their interactions with their herbivores. Though research has not yet fully unraveled what elicits these defense responses, it has clearly shown that such responses are clear and targeted. VOCs work with a plant's internal hormonal system to increase its defense. The system is not all based on altruism. There is something for everyone as long as you are close enough to pick up the signals.

In all plants studied so far, there are notable similarities in the structure of the VOCs that are emitted from insect-damaged leaves and from leaves distant from the site of damage. The uniformity in the chemical emissions from different plants with insect feeding suggests that this type of defense evolved early in plants and has been maintained because it is so effective. Its uniformity also means that these VOCs are detectable by a broad spectrum of insect parasitoids and predators. The ability of host-seeking insects to recognize and respond to such chemical cues and differentiate them from background odors indicates that insect-damaged plants emit volatile chemicals that are clearly distinguishable from those released in response to other types of damage or those released from undamaged plants. The plant's ability to differentiate between herbivore damage and a

Herbivory results in the release of volatile organic compounds (VOCs) into the air, which then communicates danger to plants nearby and causes them to respond. Photograph by Christina Evans.

general wound response suggests the presence of elicitors associated with insect feeding that are absent from other types of leaf damage. Research has shown also that these defense chemicals are not stored in the plant and then released once the damage occurs; they are specifically produced in response to the damage itself.

Not all VOC communication is directed at herbivory. Plants can emit VOCs in response to a variety of other stresses including drought and other extreme environmental conditions. In these situations, the VOCs alert the neighbors and tell them to prepare themselves appropriately. As discussed in chapter 10, plants then begin the production of certain hormones that provide some level of protection. They do this anyway when they are under stress, but communication among plants using VOCs allows them to prepare a bit in advance, and this earlier preparation can make all the difference between survival and death.

The production of VOCs and their role in above-ground plant communication may allow plants to recognize each other and distinguish between kin and strangers. Research done primarily with wild tobacco and sagebrush (*Artemisia* spp.) plants has shown that when sagebrush

Parasitic plants, such as dodder (*Cuscuta* spp.), can detect the volatile organic chemicals released by their preferred host plants and actively search and find them.

is damaged, the sensitivity of neighboring wild tobacco and other sagebrush plants differs. In a series of experiments, it was shown that only wild tobacco plants growing very close by responded to the VOCs of the injured sagebrush whereas other sagebrush responded at distances of up to six times farther away. More recent studies using sagebrush that were clones of one another showed that significantly less damage occurred to neighboring clones of a damaged plant than to its unrelated neighbors. Seemingly, plants recognize their relatives and seek to protect them via VOC-induced communication to a greater extent than they seek to protect nonrelatives. Plants are not so different from animals in this regard than we might have supposed.

Parasitic plants, such as dodder (*Cuscuta* spp.), also recognize the VOCs produced by various plants and use them to search for appropriate hosts. When given a choice between an appropriate host and an inappropriate one, dodder seedlings invariably grow toward the appropriate host and avoid the other. When blocked from detecting the VOCs, however, the dodder has no ability to make such choices, and it grows in a disoriented manner.

BELOW-GROUND METHODS

While the study of how plants communicate and what they talk about mostly began with the discovery of above-ground methods, a majority of more recent research has examined communication below ground. Below-ground communication has few of the limitations inherent with above-ground methods. VOCs are rapidly dispersed in the air and made less effective by distance and physical forces such as humidity and wind speed. Communication beneath the soil surface via the roots can seemingly remain effective for great distances from the original emitter of the signal. Root systems extend beyond the main stem, they are further enhanced by their association with mycorrhizal fungi, and these intertwine with the root systems of every other nearby plant. In a forest, for example, this circuit board of interconnected root systems provides a network that facilitates the communication of nearly every plant in the forest with each other. Underground, plants can talk about the same types of things that they do above ground, but they can do it more efficiently and effectively.

The region of the ground populated by roots, the rhizosphere, is a complex area populated by competing roots and both beneficial and harmful microbes. The ability to navigate all of this confusion lies in effective communication, and research has identified that this is done primarily through chemicals secreted by the roots. Root secretions initiate communication between plants and then direct the conversation.

As discussed in chapter 5, the root tip (i.e., meristem) produces new cells for both the root itself and the border cells that serve various functions ahead of the root's advance through the soil. Plants depend heavily on the ability of their roots, with their border cells, to communicate with microbes. All of the positive associations on which plants depend require the border cells to enlist the assistance of specific fungi, bacteria, and nematodes. Without this communication, the plant would be woefully inadequate at water and nutrient absorption as well as certain types of plant defense. The converse is also true. Many soil bacteria and fungi are dependent for their survival on their association with plants. It is a mutualistic relationship in which both sides gain.

Perhaps the most important plant communication method occurs underground, connecting adjacent root systems via mycorrhizal fungi in a grid that outrivals any computer network produced by humans.

Research has shown that parent plants in a forest can recognize their own offspring and direct resources to them while they are still finding their way up into the canopy sunlight.

changes in the secretion of various chemicals by the root system, which then causes a change in the behavior of the targeted plants. The health of a plant community is dependent on the development of a healthy underground communication network, and this is predicated on the presence of a healthy, well-developed soil. Below-ground communication serves to nurture the community and allows for the less fortunate to receive assistance from those with plenty.

Plants also use their below-ground communication network to warn each other of danger. Defense signals through this network have been shown to result in clear behavioral responses aimed at protecting its members. Even when the ability to use VOCs are blocked, plants have been shown to produce sudden changes in their foliage that results in increased pest resistance. Broad beans (*Vinca faba*) under the attack of aphids, for example, swiftly signal others via their below-ground network, and all the plants in the network then begin producing anti-aphid repellent chemicals and chemicals meant to attract ladybug beetles and other aphid predators. In another study, defoliation of Douglas fir (*Pseudotsuga menziesii*) resulted in an increased production of defense chemicals in neighboring ponderosa pines (*Pinus ponderosa*) as well.

Such communication networks develop slowly and only in healthy, stable situations. They can be easily disrupted to the detriment of the entire community. Mortality of the dominant community members through disease, pest, or intensive logging can significantly alter the mycorrhizal community beneath the soil, and this can produce a cascading effect on the trees that remain or are replanted. In areas of western North America where the mountain pine beetle has caused significant mortality of lodgepole pine (*Pinus contorta*), seedlings grown in this soil had markedly reduced growth relative to seedlings grown in noninfested stands. The disruption of the delicate mycorrhizal communication network has a legacy effect that can last for generations.

Plants also use their below-ground communication network to recognize their "selfness" from those that are not related. As should be expected, plants tend to treat their relatives with more care than those to which they are not related. Recent research has shown that when plants are grown in proximity to their relatives, root competition is purposely reduced compared with unrelated plants of the same species grown next to each other. These studies give firm support to the idea that recognition of unrelated competing plants results in an increase in root production. Recognizing one's kin and sharing resources with them would seem to favor one's long-term fitness. Just as animals recognize their close relatives and often treat them more gently, plants have evolved to behave similarly. It simply makes evolutionary sense. We are not so different from plants as we face the same long-term pressures to survive and leave healthy offspring behind.

AUDIBLE SIGNALS

Biologists have known for some time that plants communicate using VOCs and below-ground root-mycorrhizal fungi networks, though many of the exact mechanisms still remain unknown. More recently, there has emerged evidence that at least some species make auditory sounds that can be "heard" and interpreted by other plants. Auditory communication has always been in the animal realm. It would seem that at least some plants have adopted it too.

As far back as 1973, South African botanist Lyall Watson claimed that plants had emotions that could be recorded on a lie detector test. All kinds of nonscientific publications then came out proclaiming that plants responded to the sounds of their growing environment. People started talking soothingly to their house plants and played classical music to them rather than heavy metal rock and roll. The idea that plants can hear and respond to environmental sounds has largely been discounted, but recent research may show that plants are far more sensitive than we have thought.

Using powerful sound-detection equipment, researchers at the University of Western Australia discovered that corn seedlings made "clicking" sounds and that these sounds were "heard" by other corn seedlings and used as a communication tool. Researchers at nearby Bristol University then found that when exposed to the frequency of these sounds, plants grew toward them. Clearly, sound was both generated by the plants and understood at some level.

Scientists now suspect that sound and vibration may play an essential role in the survival of plants by giving them information about the environment around them. Because sound waves are easily transmissible through soil, such a system could be used to pick up threats like drought or herbi-

Recent research has discovered that some plants, such as corn (*Zea mays*), actively produce audible sounds from their root tips that are "heard" by adjacent plants and responded to.

vores from their neighbors farther away. How this system functions, however, is unknown at this time. What we do know is that plants are far more complex and interesting than anyone may have guessed.

DO PLANTS HAVE "EYES"?

As discussed in chapter 1, scientists have long known that plants can detect light and make changes to the position of their stems and flowers based on the position of the sun and/or its predicted position at sunrise. While some recent articles in popular science magazines seem to relate that to having some sense of sight, it is a far cry from demonstrating a true sense of sight.

Some recent work, however, provides evidence that plants may, in fact, possess a sense of vision that goes beyond the simple detection of light. In 1907, Charles Darwin's son Francis hypothesized that plants had organs on their leaves that were a combination of lens-type cells and light-sensitive ones. He called these "ocelli." Experiments soon followed that seemed to confirm their existence, but little came of it until recently.

In a January 2017 issue of *Trends in Plant Science* ("Plant Ocelli for Visually Guided Plant Behavior," vol. 22, no. 1), František Baluška and Stefano Mancuso provide evidence that plants are visually aware of their surroundings. They make their case by first discussing the 2016 discovery that *Synechocystis* cyanobacteria, single-celled photosynthetic organisms, act like the ocelli first postulated by Francis Darwin. Research on these cyanobacteria showed that they use their entire cell body as a lens to focus an image of the light source at the cell membrane, as in the retina of an animal eye. Although researchers are still uncertain what the purpose of this mechanism is, its existence suggests that a similar mechanism could be present in the higher plants that evolved from them.

Recent work on wild mustard has shown that it makes proteins similar to those involved in the development and functioning of eyespots in various photosynthetic bacteria. These proteins specifically show up in structures called plastoglobuli. This discovery suggests that plastoglobuli in higher plants may provide the same function.

Other observational research reveals that plants have other visual capabilities that are not yet fully understood. For example, recent studies of the climbing wood vine (*Boquila trifoliolata*) show that they can modify their leaves to mimic the colors and shapes of its host plant. Although this could be explained by communication methods other than sight, it is an intriguing explanation, given what is now known about plastoglobuli in plant leaves.

Although the evidence for eyelike structures in higher plants remains limited, it is growing, and increased attention is now being given to it within the research community. The challenge is to document the presence of plant cells that act like lenses as they do in certain cyanobacteria and then to discover what plants might use this type of vision for. If plants can actually see, how is this used to their advantage in their daily lives?

DO PLANTS HAVE "BRAINS"?

Plants do not have a central nervous system analogous to that of higher animals. A brain and spinal cord, however, are not required for an organism to exhibit intelligence. Plants can be intelligent without those organs. Modern research has shown that plants have astounding abilities to react to the world around them. They are capable of sensing nearly everything that animals do and a good number of things that animals can't. They learn from their experiences, they make complicated decisions about their futures, and they recognize their "selfness." Throughout their lives, they persevere through all the stresses the surrounding world throws at them, they assist their neighbors, and they attempt to leave their offspring better off than they would have been without their help.

Plant intelligence is remarkable because of its lack of a unifying central processor. We humans assume you need a brain to process external and internal information and make intelligent decisions regarding it. Plants show us that this assumption is not necessarily true.

Current research suggests that plants can make decisions based on past experience. Research done on sensitive mimosa demonstrates that it can change its leaf-folding behavior based on whether the trigger is benign or not. When herbivores attack it, sensitive mimosa plants continue to fold their leaves to make them look less tempting. When routinely prodded with a probe, however, they soon change their response and stop being "sensitive." What is even more amazing is that these plants remember their responses weeks later, longer than some animals might. If we look at intelligence merely as an ability to do complex problem solving, plants have to be considered intelligent.

The question remains, however, about how plants actually process this information when they don't have a brain or spinal cord. Some plant biologists argue that they must have something analogous. From as far back as 1880, Charles and Francis Darwin stated that the embryonic plant radicle not only behaves like a brain by directing the functions of other cells but also is positioned in the corresponding place in the anatomy of the plant. Some modern botanists have extended this idea. In 2005, the first international plant neurobiology meeting was held in Florence, Italy, and a brand-new journal, *Plant Signaling and Behavior*, was launched the next year. These

Research suggests that plants can make decisions based on past experience. This sensitive mimosa (*Mimosa pudica*) becomes less sensitive to touch if repeatedly touched with no harm to the plant.

new plant biologists are proposing something that would have been severely ridiculed just a decade before.

The idea that plants have some type of nervous system comes from several sources of information. First, plants have genes that are similar to those that form components of animal nervous systems. Such components include receptors for glutamate, an amino acid that, among other things, functions as a neurotransmitter. Other components are neurotransmitter pathway activators that bind various signaling proteins. These activators have been shown to have distinct roles in neural function in animals. They are also found in plants. Although most plant biologists do not believe that these proteins have "neural" functions in plants, other plant proteins seem to behave in similar ways.

What is most interesting, perhaps, is that some plants seem to show synapse-like regions between cells, across which neurotransmitter molecules facilitate cell-to-cell communication. These synapse-like regions have the same characteristics as animal synapses; they form vesicles, small bubbles that store the neurotransmitters that are to be released across the synapse. The vascular systems of most plants are structured in a way that would enable them to act as conduits for the "impulses" that they need to transmit throughout the body of the plant, and some plant cells display what could be interpreted as action potentials—events in which the electrical polarity across the cell membrane does a quick, temporary reversal, as occurs in animal neural cells.

It should not be surprising that plants share some of the same basic genes found in animals. Life on earth shares a common ancestry, and various modern genome studies verify that plants and humans share about 17 percent of our genes. It is even higher if we compare invertebrates, such as flies and worms, to plants. The question regarding the presence of glutamate remains intriguing. It has several functions in animals, some not directly involved in its action as a neurotransmitter. Its presence in plants, therefore, may have disparate functions also. It does not have to indicate neurotransmission but may have totally different functions.

When looked at closely, glutamate receptors in plants are very different from those found in animals. Plants seem to have evolved their receptors from a single common ancestor before they split from animals evolutionarily. After the split, animals seem to have taken these genes and altered them further. The receptors in animals have a one-to-one relationship to distinct organs. That is not the case with plants. Nevertheless, both groups once shared this common receptor pathway, and it has persisted in both. This then leads to the question: If glutamate receptors do not provide the same nervous system functions in plants, why have they persisted and what is their function? The most accepted argument among plant biologists is that it is involved in the communication process that stimulates the synthesis of various proteins to ward off insect attack.

Plants have maintained their glutamate receptors for a purpose. If such a purpose included functions related to neural function, there would have to be corresponding structures that behaved as neural synapses to carry information throughout the plant body. Synaptic communication would have to be demonstrated, and it would have to be performed by some type of neurotransmitter. Some plant biologists suggest that the plant hormone auxin may function in this role. There are transporter cells in plants that assist in moving auxin across cell membranes, from cell to cell. Molecular botanist Gerd Jürgens at Germany's Max Planck Institute for Developmental Biology has shown that auxin transport is accomplished through "vesicle trafficking," a process involving cellular vesicles that has animal neurotransmitter-like features. Some plant biologists believe that this process is similar enough to the way animals transmit information to link this process to a neurological function.

If plants have a neurological transmitter and a pathway, do they also create electrical impulses along that pathway similar to the process shown in animals? The answer is yes. Electrical conductivity in plants was actually demonstrated several years before it was shown in animals. Its exact function, however, has only recently been studied. While plants and animals have evolved very different systems, each has a need for information to travel from a point of disturbance to the rest of the body and for decisions to be made regarding an appropriate response. Whereas animals produce the action potential by an exchange of sodium and potassium ions, plant potentials are produced with calcium transport that is enhanced by chloride and reduced by potassium.

So, do plants have "brains" and a definable nervous system? That remains the most controversial aspect of modern plant behavioral research. Some, including Anthony Trewavas of the University of Edinburgh, suggest that plant neurobiology is nothing more than a metaphor. Such biologists believe that there are no corresponding neurological structures in plants, and their research seeks to redefine the concept of neurobiology for plants in the context of plant cell-to-cell communication and signaling. Other biologists, however, are not willing to discount the existence of an actual nervous system in plants.

Animal nervous systems arose as a conducting device tailored to specific constraints related to coordinating free-moving behavior. Plants are also highly evolved multicellular beings that must coordinate their behavior in responses to a wide variety of internal and external signals. One would expect that plants evolved their own conductive devices for this purpose. In particular, when the huge size of some plants is taken into account, the presence of long-distance signaling seems to be essential, and the functional need for a nervous system—or at least an analogous neural system—seems obvious.

Where, then, would a plant "brain" be located? The focus of this work is being directed at the root apical meristem. As discussed above, biologists continue to discern the very real complexities of this region of the plant, including its role in plant communication and the production and role of border cells. This seems to be the logical location for a plant "brain."

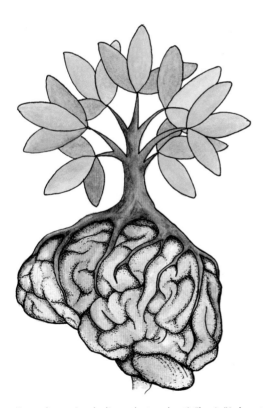

Some botanists believe that a plant's "brain" is located in its roots.

The transition zone of the root apical meristem plays a critical role in the circuitry of auxin transport and the integration of different types of signals. Cells in the transition zone change polarity in response to glutamate activity, and the cell vesicles are then able to transport auxin throughout the entire plant. The transition zone is special. It is the one and only plant area where electrical activity is known to synchronize and where brain-like decision making takes place that affects every plant organ. Whether this suggests an actual brain center and a defined neurological pathway remains controversial. It does, however, seem to set a clear stage that allows for future comparative studies between plants and animals.

Further research will likely redefine the way we look at plants and understand them. If anything, it is likely to remove, once and for all, the schism between the plant and animal world that has existed since Aristotle.

Conclusion

Life itself is a miracle, regardless of how it is manifested. As humans, we often encumber our concept of life with value judgments, giving some forms higher value than others. Since the birth of our widespread Eurocentric view of science, plants have been considered to occupy one of the lowest links on what Joseph Wood Krutch called the "great chain of life." They seem uncomplicated and slow, and if they didn't provide us the services of food, clothing, and shelter, we might continue to ignore them as Paleolithic peoples seemingly did thousands of years ago.

The reality is that plants live within their own universe, almost parallel to ours, a universe that is both dazzling and complicated. Its existence is within a different time scale than ours, visible only on close inspection by those interested enough to look. Plants live their lives just as we do, surrounded by loved ones, competitors, and enemies; seeking to find harmony and health; and hoping to leave behind a legacy of well-adjusted progeny capable of carrying on after their demise. Like us, they have evolved complicated pathways and behaviors to accomplish this. We are not really that different, and different should not imply a value judgment. It is merely a statement of fact.

Thank you for exploring the world of plants with me. Perhaps your understanding of plants and how they work has been changed a bit. If we have done that together, we will have accomplished something significant both for ourselves and this amazing world we share with the rest of creation. Tend your plants and garden with reverence.

Index

Page numbers in **bold** indicate a photograph or illustration.

CRAIG N. HUEGEL is owner and operator of Hawthorn Hill Native Wildflowers. He is a former faculty member of the Wildlife Ecology and Conservation department at the University of Florida, where he cofounded the Cooperative Urban Wildlife Extension Program. He is the author of *Native Florida Plants for Shady Landscapes*, *Native Wildflowers and Other Ground Covers for Florida Landscapes*, and *Native Plant Landscaping for Florida Wildlife*.